Li Niu, Jie Lu, and Guangquan Zhang

Cognition-Driven Decision Support for Business Intelligence

Studies in Computational Intelligence, Volume 238

Editor-in-Chief
Prof. Janusz Kacprzyk
Systems Research Institute
Polish Academy of Sciences
ul. Newelska 6
01-447 Warsaw
Poland
E-mail: kacprzyk@ibspan.waw.pl

Further volumes of this series can be found on our homepage: springer.com

Vol. 216. Matthew Taylor
Transfer in Reinforcement Learning Domains, 2009
ISBN 978-3-642-01881-7

Vol. 217. Horia-Nicolai Teodorescu, Junzo Watada, and Lakhmi C. Jain (Eds.)
Intelligent Systems and Technologies, 2009
ISBN 978-3-642-01884-8

Vol. 218. Maria do Carmo Nicoletti and Lakhmi C. Jain (Eds.)
Computational Intelligence Techniques for Bioprocess Modelling, Supervision and Control, 2009
ISBN 978-3-642-01887-9

Vol. 219. Maja Hadzic, Elizabeth Chang, Pornpit Wongthongtham, and Tharam Dillon
Ontology-Based Multi-Agent Systems, 2009
ISBN 978-3-642-01903-6

Vol. 220. Bettina Berendt, Dunja Mladenic, Marco de de Gemmis, Giovanni Semeraro, Myra Spiliopoulou, Gerd Stumme, Vojtech Svatek, and Filip Zelezny (Eds.)
Knowledge Discovery Enhanced with Semantic and Social Information, 2009
ISBN 978-3-642-01890-9

Vol. 221. Tassilo Pellegrini, Sören Auer, Klaus Tochtermann, and Sebastian Schaffert (Eds.)
Networked Knowledge - Networked Media, 2009
ISBN 978-3-642-02183-1

Vol. 222. Elisabeth Rakus-Andersson, Ronald R. Yager, Nikhil Ichalkaranje, and Lakhmi C. Jain (Eds.)
Recent Advances in Decision Making, 2009
ISBN 978-3-642-02186-2

Vol. 223. Zbigniew W. Ras and Agnieszka Dardzinska (Eds.)
Advances in Data Management, 2009
ISBN 978-3-642-02189-3

Vol. 224. Amandeep S. Sidhu and Tharam S. Dillon (Eds.)
Biomedical Data and Applications, 2009
ISBN 978-3-642-02192-3

Vol. 225. Danuta Zakrzewska, Ernestina Menasalvas, and Liliana Byczkowska-Lipinska (Eds.)
Methods and Supporting Technologies for Data Analysis, 2009
ISBN 978-3-642-02195-4

Vol. 226. Ernesto Damiani, Jechang Jeong, Robert J. Howlett, and Lakhmi C. Jain (Eds.)
New Directions in Intelligent Interactive Multimedia Systems and Services - 2, 2009
ISBN 978-3-642-02936-3

Vol. 227. Jeng-Shyang Pan, Hsiang-Cheh Huang, and Lakhmi C. Jain (Eds.)
Information Hiding and Applications, 2009
ISBN 978-3-642-02334-7

Vol. 228. Lidia Ogiela and Marek R. Ogiela
Cognitive Techniques in Visual Data Interpretation, 2009
ISBN 978-3-642-02692-8

Vol. 229. Giovanna Castellano, Lakhmi C. Jain, and Anna Maria Fanelli (Eds.)
Web Personalization in Intelligent Environments, 2009
ISBN 978-3-642-02793-2

Vol. 230. Uday K. Chakraborty (Ed.)
Computational Intelligence in Flow Shop and Job Shop Scheduling, 2009
ISBN 978-3-642-02835-9

Vol. 231. Mislav Grgic, Kresimir Delac, and Mohammed Ghanbari (Eds.)
Recent Advances in Multimedia Signal Processing and Communications, 2009
ISBN 978-3-642-02899-1

Vol. 232. Feng-Hsing Wang, Jeng-Shyang Pan, and Lakhmi C. Jain
Innovations in Digital Watermarking Techniques, 2009
ISBN 978-3-642-03186-1

Vol. 233. Takayuki Ito, Minjie Zhang, Valentin Robu, Shaheen Fatima, and Tokuro Matsuo (Eds.)
Advances in Agent-Based Complex Automated Negotiations, 2009
ISBN 978-3-642-03189-2

Vol. 234. Aruna Chakraborty and Amit Konar
Emotional Intelligence, 2009
ISBN 978-3-540-68606-4

Vol. 235. Reiner Onken and Axel Schulte
System-Ergonomic Design of Cognitive Automation, 2009
ISBN 978-3-642-03134-2

Vol. 236. Natalio Krasnogor, Belén Melián-Batista, José A. Moreno-Pérez, J. Marcos Moreno-Vega, and David Pelta (Eds.)
Nature Inspired Cooperative Strategies for Optimization (NICSO 2008), 2009
ISBN 978-3-642-03210-3

Vol. 237. George A. Papadopoulos and Costin Badica (Eds.)
Intelligent Distributed Computing III, 2009
ISBN 978-3-642-03213-4

Vol. 238. Li Niu, Jie Lu, and Guangquan Zhang
Cognition-Driven Decision Support for Business Intelligence, 2009
ISBN 978-3-642-03207-3

Li Niu, Jie Lu, and Guangquan Zhang

Cognition-Driven Decision Support for Business Intelligence

Models, Techniques, Systems and Applications

Springer

Li Niu
University of Technology Sydney (UTS)
Fac. Information Technology
PO BOX 123
Broadway NSW 2007
Australia
E-mail: ollien@gmx.com

Guangquan Zhang
University of Technology Sydney (UTS)
Fac. Information Technology
PO BOX 123
Broadway NSW 2007
Australia
E-mail: zhangg@it.uts.edu.au

Jie Lu
University of Technology Sydney (UTS)
Fac. Information Technology
PO BOX 123
Broadway NSW 2007
Australia
E-mail: jielu@it.uts.edu.au

ISBN 978-3-642-03207-3 e-ISBN 978-3-642-03208-0

DOI 10.1007/978-3-642-03208-0

Studies in Computational Intelligence ISSN 1860-949X

Library of Congress Control Number: Applied for

Typeset & Cover Design: Scientific Publishing Services Pvt. Ltd., Chennai, India.

Printed on acid-free paper

9 8 7 6 5 4 3 2 1

springer.com

Preface

Cognition-driven decision support system (DSS) has been recognized as a paradigm in the research and development of business intelligence (BI). Cognitive decision support aims to help managers in their decision making from human cognitive aspects, such as thinking, sensing, understanding and predicting, and fully reuse their experience. Among these cognitive aspects, decision makers' situation awareness (SA) and mental models are considered to be two important prerequisites for decision making, particularly in ill-structured and dynamic decision situations with uncertainties, time pressure and high personal stake. In today's business domain, decision making is becoming increasingly complex. To make a successful decision, managers' SA about their business environments becomes a critical factor.

This book presents theoretical models as well practical techniques of cognition-driven DSS. It first introduces some important concepts of cognition orientation in decision making process and some techniques in related research areas including DSS, data warehouse and BI, offering readers a preliminary for moving forward in this book. It then proposes a cognition-driven decision process (CDDP) model which incorporates SA and experience (mental models) as its central components. The goal of the CDDP model is to facilitate cognitive decision support to managers on the basis of BI systems. It also presents relevant techniques developed to support the implementation of the CDDP model in a BI environment. Key issues addressed of a typical business decision cycle in the CDDP model include:

- natural language interface for a manager's SA input;
- extraction of SA semantics;
- construction of data warehouse queries based on the manger's SA and experience;
- situation information retrieval from data warehouse;
- how the manager perceives situation information and update SA;
- how the manager's SA leads to a final decision.

Finally, a cognition-driven DSS, FACETS, and two illustrative applications of this system are discussed.

Two important features of this book clearly distinguish itself from other books in similar areas:

(1) It is the first book to systematically discuss the theories, technologies and applications of cognitive decision support for BI.

(2) It reflects the latest academic research progress as well as the state-of-art BI technologies by combing contemporary cognitive psychology, DSS, BI, and data warehousing.

This book is mainly based on our past few years' research developments in this field. The technologies presented in this book is up-to-date throughout some results come from ours and other authors' recent publications.

The potential readers of this book are organizational managers and practicing professionals, who can use the provided methods and software to solve their real decision problems; researchers in the areas of decision making, DSS and BI; students at the advanced undergraduate or Master's level in management or business administration programs; and students at the advanced undergraduate or Master's level in information systems, information technology and computer science programs.

This book was organized into four major parts.

(1) Concept part (Chapters 1, 2, 3 & 4) covers concepts and frameworks of decision making and general decision-making techniques in DSS.

(2) Model part (Chapter 5) presents readers with the framework of cognitive decision support and the model of cognition-driven decision processes.

(3) Technique part (Chapters 6, 7 & 8) discusses the specific techniques (methods and algorithms) of cognitive decision support. Readers will learn how information technologies are combined with cognitive psychology models to solve business decision-making problems.

(4) System and Application part (Chapters 9, 10 & 11) introduces a cognition-driven DSS, FACETS. FACETS is a successful implementation of the cognition-driven decision process model, which was developed based on the models and techniques presented in this book. This part also demonstrates the evaluation results of FACETS and its two illustrative applications in business and public health respectively.

Readers can gain various advantages from this book:

As an academic, you will know the latest research progress of cognitive DSS in our research laboratory. This book provides a wide range of literature in many related areas, such as BI, data warehousing, cognitive psychology, naturalistic decision making, recognition-primed decision making, and natural language interface to databases. It also points out some potential topics for further research.

As an information technology practitioner, reading this book, you will stay abreast of BI, and knowing the new generation of BI systems. The models, techniques and algorithms presented in this book will help you understand how to better support business managers' work using the state-of-art BI technology.

As a business manager, you will be presented with a new methodology to interpret your business problems from a new perspective. The technologies included in this book will provide a new vision for you to understand and handle your business problems. After reading the book, you should be able to identify decision-making problems from real practices, to build mental models, and then to use the cognition-driven DSS provided in the book to support your decision making.

As a tertiary student, you will have a chance to touch on a wide rang of information technologies, which will be a starting point for your future study and career. You will also learn how to convert business problems into research projects.

We wish to thank the Australian Research Council (ARC) as the work presented in this book was partially supported under ARC Discovery Projects DP0559213 and DP0880739; the many researchers who have worked in DSS, BI and related areas over the past several decades, whose significant insights we have drawn on in the book and whose well-known publications are included in the bibliography; the researchers and students in the Decision Systems & e-Service Intelligence (DESI) research laboratory, and from the Faculty of Engineering and Information Technology at the University of Technology Sydney (UTS), who suffered through several versions of the DSS presented in this book and whose comments improved it substantially; and the editors and production staff at Springer, who helped us to ensure the book was as good as we were capable of making it.

Sydney
May 2009

Li Niu,
DESI, University of Technology, Sydney

Jie Lu,
DESI, University of Technology, Sydney

Guangquan Zhang,
DESI, University of Technology, Sydney

Contents

Part I: Concepts

Part II: Models

Part III: Techniques

Part I
Concepts

Chapter 1
Decision Making and Decision Support Systems

This chapter introduces the general concepts, models and techniques of decision making and decision support systems. It also gives readers an overview of the new models and techniques presented in the following chapters of this book.

1.1 Decision Making and Decision Makers

The success of business management depends on the performance of managerial functions such as planning, organizing, directing, and controlling (Turban et al. 2005). To carry out these functions, business managers are engaged in a continuous process of making decisions such as drawing up a product plan, selecting a supplier and determining a product's price. In such situations, business managers are decision makers.

Decision makers may deal with various types of decision problems, from daily operation to long-term company strategies. Decision makers in a company act at various levels, from a software development project manager to a CEO of a large company. Therefore, different decision-making tasks have different features and require different decision support techniques.

Decisions can be made by individuals or groups. Individual decisions are often made at low management levels and in small organizations, while group decisions are usually made at high management levels and in large businesses. In group decision making, each group member has their own understanding of the problem and the solution. Group members participate in decision making with different capabilities and action resources. Group decision making is more complex than individual decision making, due to the conflicts between different decision makers' self-interests and preferences. Therefore, communication and interaction among group members are very important in group decision making.

Decision making theory is studded by many researchers from multiple disciplines (Dawes 1988; Hwang & Masud 1979; Simon 1979). It includes classical decision making (CDM), behavioural decision theory (BDT), judgement and decision making (JDM), organizational decision making (ODM) and recently naturalistic decision making (NDM) (Lipshitz et al. 2001). Choice and input-output orientation are the examples of CDM. Some new theories such as the use of

L. Niu, J. Lu, and G. Zhang: Cognition-Driven Dec. Supp. for Business Intel., SCI 238, pp. 3–18.
springerlink.com

expertise in seizing up situations and generating options are examples of NDM. Decision making is growing increasingly complicated and difficult today due to information overload and fluctuated decision environments. With the modern information technologies and communication systems, we can find large quantities of information quickly and easily, and therefore generate more alternatives. Nevertheless, the changing decision environment impose decision making more uncertainties, which requires dynamic decision making and needs decisions to be made quickly. The cost of making errors can be very large because of the complexity of operations, automation, and the chain reaction that an error can cause in many parts of a business. For these reasons, business decision makers require technical support to help make quality decisions in a short timeframe.

1.2 Decision Problem Classification

Decision problems can be classified according to their natures. One important classification is based on a given problem structure: *structured, semi-structured* and *unstructured* (Turban et al. 2005). We use the term *ill-structured decision problems* to describe both semi-structured and unstructured decision problems. Different classes of decision problems require different modeling methods.

A *structured decision* problem can be described by existing classic mathematical models, such as linear programming and statistics methods. The procedures for obtaining the best or the most satisfactory solution are known as standard solution methods. Examples of typical structured decision problems are selecting a product which has the lowest price among all the same type of products, and determining which product plan can bring the highest profit among a range of product plans.

An *unstructured decision* problem is fuzzy, uncertain and vague, to which there is no standard solution method. Human intuition is often the basis for decision making in an unstructured problem. Typical unstructured problems include planning new services, hiring an executive, and choosing a set of development projects for a long period.

Semi-structured decision problems fall between structured and unstructured problems, having both structured and unstructured factors. Solving this kind of decision problems involves a combination of both standard optimized solution procedures and human intuition or judgments.

Experience shows that computer-based decision support techniques are more useful in structured decision problems than in semi-structured and unstructured decision problems. In an unstructured decision problem only part of the problem can be well-supported by computerized decision support tools. For semi-structured decision problems, computerized support technologies can improve the quality of the information on which the decision is based, increase decision makers' situation awareness, or provide not only a single solution but a range of alternative solutions.

Another classification of decision problems is based on decision levels as proposed by Anthony (1965): *strategic planning, management control* and *operational control.*

Strategic planning refers to long-range goals and policies for resource allocation. Such decisions are at a high management level, normally unstructured, and with higher degrees of uncertainty.

Management control refers to the acquisition and efficient use of resources in the accomplishment of business goals, and related decisions are at a middle management level.

Operational control decisions are about the efficient and effective execution of specific tasks. They are normally structured and are relatively easy to formulate by mathematical models and solved using computer-based tools.

Decision making is a reasoning process, rational or irrational (Simon 1979, Simon 1993), and can be based on explicit or tacit assumptions.

Rational decision making emphasizes fact collection and conducting research such as data analysis, surveys and interviews. A rational decision-making model involves a cognitive process (thinking through) where each step follows in a logical order from the one before.

Irrational decision making makes assumptions and obtains results without accurate data and model analysis, and is often driven by emotions.

1.3 Decision-Making Process

Generally speaking, a decision-making process (or simply called decision process) begins with awareness of a decision problem, and ends up with a final solution among finite or infinite alternatives. A general decision-making process proposed by Simon (1977) involves three phases: *Intelligence*, *Design*, and *Choice*. A fourth phase, *Implementation*, has since been added. Figure 1.1 shows a conceptual framework of the four-phase decision-making process. This framework can be used as a guideline for specific decision-making development. Different decision makers may emphasize one phase or another. Different decision-making problems may require more details or sub-phases in one or more phases. We will discuss an extended decision-making process model (Lu et al. 2007) based on Simon (1997) as shown in Figure 1.1.

The model shown in Figure 1.1 is the most typical decision-making process model. Its main characteristic is to establish a decision model to reflect a decision problem and then to solve the problem by applying a method to the decision model. The decision-making process can be described by nine steps. We use an example to explain these steps as follows:

Example 1.1: A logistics company plans to develop an e-business system.

Step 1: Define a decision problem

To define a decision problem requires a good understanding of managerial assumptions, organizational boundaries, and any related initial and desired conditions. This step aims to express the decision problem in a clear way and prepare a clear *problem statement* (see Figure 1.1). This step, with Step 2, corresponds to the *intelligent phase* of the general decision-making process framework.

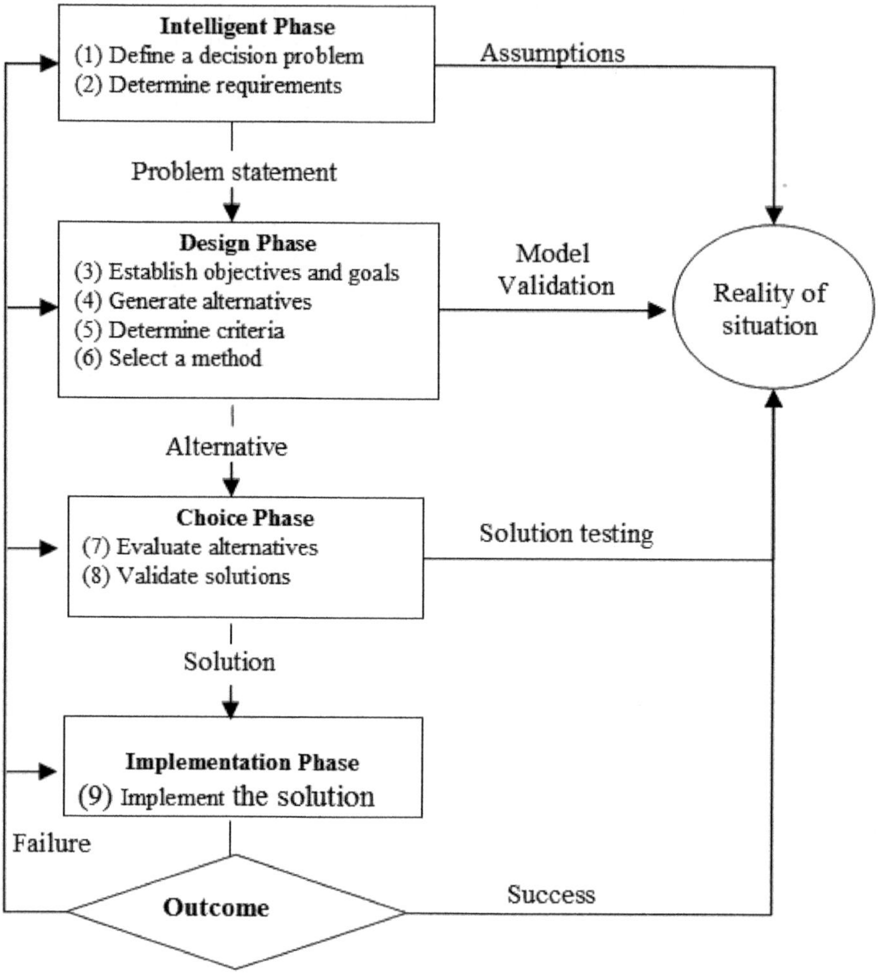

Fig. 1.1 The general decision-making process framework

The decision problem defined in this step for Example 1.1 is

"select an IT company for developing an e-business system for the logistics company".

Step 2: Determine requirements
Requirements are conditions which any acceptable solution to the problem must meet. In a mathematical form, these requirements are the constraints describing the set of the feasible solutions of the decision problem such as $cost1+cost2<\$100$. Requirements can be obtained by analysing the decision situation.

The requirements determined for Example 1.1 are

"the cost of the e-business system development is to be <=$100,000, the deadline is 30 Dec 2009, and the developed e-business system must connect well with existing business information systems and data warehouse".

Step 3: Establish objectives and goals
The design phase of the decision-making process starts here and continues through to Step 6. This step identifies the important objectives of the decision problem and the goals. When a decision problem involves multiple objectives, these objectives may be in conflict with each other. The goals are the statements of intent and desirable programmatic values. In a mathematical form, the goals are objectives contrary to the requirements that are constraints. Obviously, these objectives can have different degrees of importance.

The objectives for Example 1.1 are

"find a satisfactory IT company which meets the requirements for development of an e-business system and provides an excellent user interface and all required functions".

Step 4: Generate alternatives
The objectives obtained in Step 3 are used to generate alternatives. Each alternative must meet the requirements determined in Step 2. For finite number of the possible alternatives, they can be checked one by one against the requirements. The infeasible alternatives will be removed out of further consideration until we obtain the explicit list of alternatives. If the number of possible alternatives is infinite, the set of alternatives is considered to be that of the solutions fulfilling the constraints in the mathematical form of the requirements.

The alternatives generated in Example 1.1 are

"from all the IT companies having interest in the development of the e-business system, three companies (A, B, C) are selected as alternatives based on the defined requirements and objectives".

Step 5: Determine criteria if needed
To choose the best alternative, all alternatives need to be evaluated against the objectives. Thus some criteria are used to compare alternatives and to discriminate between alternatives, based on the objectives. It is necessary to define discriminating criteria as objective measures to assess each alternative.

The criteria determined for Example 1.1 are

"the e-business system to be developed should be user-friendly, secure, and easy to maintain, and have excellent functions for logistics decision support, and ensure that the budget/cost is relatively low".

Step 6: Select a decision-making method
In general, there are always several methods or tools available for solving a particular decision problem. The selection of an appropriate method or tool

depends on the characteristics of the decision problem and the preferences of the decision maker.

The decision-making method selected for Example 1.1 is

"as the decision is based on multiple criteria and will be made by a committee, an Analytic Hierarchy Process (AHP) method will be used for aggregating different members' opinions under all criteria".

Step 7: Evaluate alternatives
This step corresponds to Simon's choice phase of decision making. After applying the determined criteria supported by the selected method, a tentative decision is made in this step through the evaluation of the alternatives against the objectives. Using a commonly shared and understood scale of measurement and the subjective assessment of the evaluation, the selected decision-making tool can be applied to rank the alternatives or to choose a subset of the most promising alternatives.

The result of alternative evaluation for Example 1.1 is

"by applying the AHP method, IT company B is chosen to undertake the development of the e-business system for the logistics company".

Step 8: Validate solutions
The tentative alternative generated in Step 7 has to be validated against the requirements and objectives of the decision problem, in order to determine its applicability. Many factors might result in an inappropriate or false choice, such as misapplied decision-making method or wrongly determined criteria. If the tentatively chosen alternative has no significant adverse consequences, the choice is finally made.

The result of solution validation for Example 1.1 is

"IT company B is accepted by all committee members".

Step 9: Implement the solution
The final solution to the identified decision problem is implemented in this step.

The result of solution implementation for Example 1.1 is

"sign a project contract with IT Company B".

As can be seen from above process, decision making is a choice process among alternatives. Each decision-making task can be characterized by a problem statement, a set of alternatives, and related decision criteria. Decision makers go through all these phases and eventually reach a final solution.

1.4 Decision Support Systems

Decision making, by its nature, is a cognitive process, involving different cognitive tasks, such as collecting information, evaluating situation, generating and selecting alternatives, and implementing solutions. Decision making is never error-proof, as decision makers are prone to their cognitive biases. Therefore, decision support systems (DSS) are often used by decision makers in order to minimize their cognitive errors and maximize the performance of actions.

1.4.1 The Concept

A decision support system is a computerized information system, designed to support business and organizational decision-making activities. The term 'DSS' was proposed in the early 1970s. After that, the topic of DSS has stimulated great interests in research and its applications. Gorry and Scott-Morton (1971) defined DSS as interactive computer-based systems which help decision makers utilize *data* and *models* to solve ill-structured problems. A subsequent classic definition of DSS, provided by Keen and Scott-Morton (1978), is that DSS couple the intellectual resources of individuals with the capabilities of the computer to improve the quality of decisions.

A properly-designed DSS can play an important role in compiling useful information from raw data, documents, personal knowledge, and business models to solve problems. It allows decision makers to perform large numbers of computations very quickly. Therefore advanced models can be supported by DSS to solve complex decision problems, e.g., emergency situations, where quick responses are often required. As many business decision problems involve large data sets stored in different databases, data warehouses, and even possibly at websites outside an organization, DSS can retrieve, process and utilize data efficiently to assist decision making.

A DSS is intended to support, rather than replace, decision maker's role in solving problems. Decision makers' capabilities are extended through using DSS, particularly in ill-structured decision situations. In this case, a satisfied solution, instead of the optimal one, may be the goal of decision making. Solving ill-structured problems often relies on repeated interactions between the decision maker and the DSS.

1.4.2 Characteristics

The functions and characteristics of DSS vary significantly due to the differences of application domains. Turban and Aronson (1998) summarized ten characteristics of common DSS as follows:

(1) Dealing with ill-structured decision problems;
(2) Supporting managers at different levels;
(3) Supporting decision groups and individual decision makers;
(4) Supporting a variety of decision styles and processes;

(5) Adaptability and flexibility in carrying out a decision support task and approach of the users;

(6) Interactive and user-friendly to allow non-technical decision makers to interact easily with it;

(7) Combining the use of models and analytic techniques;

(8) Combining the use of artificial intelligence and knowledge base;

(9) Accessing a wide variety of data sources; and

(10) Integration and Web connection.

The selection of above functions depends on a user's requirements. A DSS can be quite simple, e.g., a spreadsheet, or extremely complex, e.g., a data warehouse system.

1.4.3 Types

According to different criteria, DSS can be categorized into various types, such as personal DSS, group DSS, executive support systems, web-based DSS, desktop DSS, strategic DSS and financial planning DSS (Arnott & Pervan 2005). Based on the model of assistance, Power (2002) differentiates model-driven DSS, communication-driven DSS, data-driven DSS, document-driven DSS and knowledge-driven DSS as follows.

(1) A model-driven DSS emphasizes creation and manipulation of statistical, financial, optimization, or simulation models (Power 2002). The linear programming based optimization model is one of the most wide-used DSS models. Model-driven DSS require users (decision makers) to specify model parameters according to their decision problems. The outputs of the model are used to assist decision makers in assessing their decision alternatives. Multi-objective DSS (MODSS) are a typical model-driven DSS where multi-objective decision making models are adopted (Lu et al. 2007).

(2) A communication-driven DSS supports decision making within a group of decision makers through facilitating efficient information exchange (Power 2002). It is also called group DSS (GDSS). Information exchange stimulates intelligence development and integration, which promotes decisions to be made based on the consensus among different decision makers. Examples of this type of DSS are email systems, electronic meeting systems and bulletin board systems.

(3) A data-driven DSS focuses on access to and manipulation of a large amount of company data from internal and external sources (Power 2002). Decision making is based on perceiving and comprehending the integrated information output by the system. Examples of this type of DSS are statistics tools, management information systems and BI systems.

(4) A document-driven DSS is concerned with managing and manipulating unstructured information in a variety of electronic files, such as emails and reports (Power 2002). Examples of this type of DSS are library management systems, search engines and document retrieval systems.

(5) A knowledge-driven DSS generates decision suggestions based on human expertise (knowledge) (Power 2002). Human expertise is specific to application domains, which can be elicited from experts. The common forms of knowledge are business procedures, rule and facts. Examples of this type of DSS are case-based reasoning systems, expert systems and question-answer systems.

Due to the diversity of real applications, there is no general software architecture for DSS in different domains. Thus, the components and functions of DSS vary greatly. For example, a mathematical programming model is the core component of a model-driven DSS, while a BI system consists of a data warehouse subsystem and a data analysis subsystem. The software architecture and functions of a DSS are often designed based on users' specific requirements.

A specific DSS is usually a combination of some of the above five different types. However, the model-driven DSS dominates the traditional DSS research and applications (Arnott & Pervan 2005). As a result, many practical DSS were developed with the inclusion of a model component, although they also support other DSS functions, e.g., group communication and data-intensive analytics. Thus, the decision processes in traditional DSS are essentially model-based (Courtney 2001).

1.5 Decision Support Techniques

Decision support systems are built upon various decision support techniques, including models, methods, algorithms and tools. Zachary (1986, 1987) proposed a cognition-based taxonomy for decision support techniques, including six basic classes as follows.

Process models are computational models that assist the projection of real-world complex processes and give assumptions about the process and a hypothetical decision. A typical process model is probabilistic models which compute a probability distribution of outcomes from a probability distribution of input conditions through an analytical treatment of inputs and the behavior of the process. Markov chains are a common example of a probabilistic process model.

Choice models support the integration of decision criteria across alternatives to select the best alternative from a discrete set or continuous description space of decision alternatives. A typical choice model is the multi-criteria decision-making model.

Information control techniques provide functions of representation, manipulation, access, and monitoring data and knowledge. Typical techniques include database management tools, data and knowledge retrieval techniques, data warehousing, data mining and automatic aggregation.

Analysis and reasoning techniques support applications of problem-specific expert reasoning procedures, such as mathematical programming, goal-driven inference, process-driven inference and data-driven inference. Goal programming, evidential reasoning, case-based reasoning and sensitivity analysis are successful analysis and reasoning techniques.

Representation aids support the expression and manipulation of a specific representation of a decision problem. Typical techniques of this type include natural language processing, graphic user interface, and human cognitive processing techniques. Some examples are decision trees, decision tables, and cognitive mapping.

Human judgment amplifying/refining techniques assist decision makers in quantification of heuristic judgments. "Decision makers are able to solve problems heuristically or intuitively with results that are usually quite good but almost never truly optimal" (Zachary 1986). Typical techniques in this class include human-aided optimization, adaptive user modeling and prediction, as well as Bayesian updating.

The cognition-based classification of decision support techniques provides a picture and guideline for decision technique selection for problem solving and DSS development. In practice, a DSS often uses two or more of the techniques mentioned above to solve a problem. We will introduce five popular decision support techniques under the above taxonomy.

1.5.1 Optimization

Optimization, also called *mathematical programming,* refers to the study of decision problems in which one seeks to minimize or maximize a function by systematically choosing the values of variables from an allowed set. A mathematical programming model includes three sets of elements: *decision variables*, *objective function*(*s*), and *constraint*(*s*), where *uncontrollable variables* or *parameters* are within the objective functions and constraints. Many real-world decision problems can be modeled by a mathematical programming model. There are many types of mathematical programming models such as linear programming (Benayoun et al. 1971), multi-objective programming (Hwang & Masud 1979), and bi-level/multi-level programming (Anandalingam & Friesz 1992).

Linear programming is an important type of optimization in which the objective function and all constraints are linear. There are pure linear programming problems corresponding to specialized solution algorithms. There are also other types of optimization problems including a linear programming problem as a sub-problem. Linear programming is heavily used in various management activities, either to maximize profit or minimize cost.

1.5.2 Multiple Criteria Decision Making

Multi-criteria decision making (MCDM), also called multi-attribute decision making (MADM), refers to making preference decisions (*e.g.*, evaluation, prioritization, and selection) in the presence of multiple and conflicting criteria over available alternatives. A MCDM utility model combines all criteria of a given alternative simultaneously through a specific utility formula or utility function used. Problems for MCDM may range from those in our daily life, such as the purchase of a car, to those affecting entire nations, such as the judicious use of money for the preservation of national security. However, even with such diversity, all MCDM problems share the following common characteristics (Hwang & Yoon 1981):

A set of alternatives: there are usually a limited number of predetermined alternatives, such as three IT companies available to develop an e-business system for a logistics company.

Multiple criteria: each problem has multiple criteria. For example, development cost, quality of interface, and quality of security are criteria for selecting an IT company.

Conflicting criteria: there are multiple criteria conflicting with each other, For example, the criterion "low cost" and the criterion "high quality of security function" conflict.

Incommensurable unit: criteria may have different units of measurement. For example, the unit of cost is the dollar, while that of interface design is the degree of satisfaction.

Selection: the solution to an MCDM problem is to select the best one among previously specified finite alternatives.

Mathematically, a MCDM problem can be modeled as follows:

$$\begin{cases} \text{Select}: A_1, A_2, \ldots, A_m \\ \text{s.t.}: C_1, C_2, \ldots, C_n \end{cases}$$

where $A = (A_1, A_2, \ldots, A_m)$ denotes m alternatives, and $C = (C_1, C_2, \ldots, C_n)$ represents n criteria (or attributes) for characterizing a decision situation. The *select* here is normally based on maximizing a multi-criteria value (or utility) function elicited from the decision makers. By mapping the alternatives onto a cardinal scale of value, the alternative with the highest cardinality is implicitly the best.

Multi-criteria decision-making methods have been widely developed, as reported by Hwang and Yoon (1981) as well as Yager (2004b) and many other researchers. Some popular MCDM methods include the TOPSIS method (Hwang & Yoon 1981) and the Analytic Hierarchy Process (AHP) method (Saaty 1980).

1.5.3 Data Mining

Data mining is a data processing technique to extract hidden patterns from data of interest, in order to provide decision makers 'knowledge' for decision making. A data mining project consists of six typical stages (*www.crisp-dm.org*), which is shown in Figure 1.2.

Stage 1. Business Understanding
A data mining project begins with collecting and understanding business requirements and converting the business requirements into technical specifications.

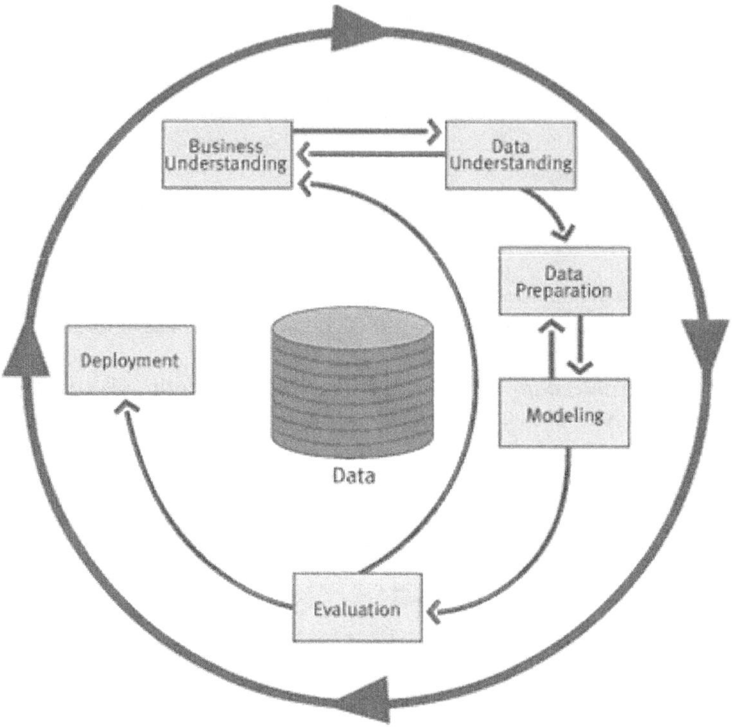

Fig. 1.2 Phases of the CRISP-DM Process Model (www.crisp-dm.org)

Stage 2. Data Understanding
The data is initially analyzed in this stage, in order to detect the problems of data quality and prospective patterns to be exposed.

Stage 3. Data Preparation
Data is pre-processed to reduce or eliminate data quality problems, such as inconsistency and data missing.

Stage 4. Modeling
Specific data modeling methods are selected and applied against the cleaned data, such as clustering, regression and classification.

Stage 5. Evaluation
The created data models are evaluated in this stage from both business and technical perspectives.

Stage 6. Deployment
If the data models are considered as good enough for the business requirements, they are deployed in the business environment to assist decision making.

Data mining is usually integrated with a decision support system. A common practice is to develop the data mining application on the basis of a data warehouse system. A data warehouse is a subject oriented, time-variant, non-volatile and integrated data store (Inmon 1993). A data warehouse system includes not only data storage, but also the techniques to extract, transform and load data, and retrieve and analyze data, and manage the data dictionary.

1.5.4 Case-Based Reasoning

Fully-automated expert systems have been used as a type of decision support system to produce a solution for a given problem statement. Such systems have been highly successful in solving problems in many well-circumscribed domains. However, they have not been successful in support decision making which requires creativity, broad commonsense knowledge, or aesthetic judgment (Kolodneer 1991).

Case-based reasoning (CBR) provides a methodology for decision support systems in solving new problems based on the solution of similar past problems. The core technique of CBR is to have a powerful learning ability which can use past experiences as a basis for dealing with novel problems. A CBR system can therefore facilitate the knowledge acquisition process by eliminating the time required to elicit solutions from experts. In dynamically changing situations where much is unknown and solutions are not clear cut, CBR seems to be the preferred method of reasoning.

Case-based reasoning is represented by a four-stage (4Rs) cycle: *retrieve, reuse, revise* and *retain* (Aamodt & Plaza 1994). In the first 'R' stage, when a new problem is input, CBR retrieves the most similar case from the case base. In the second 'R' stage, the solution of the retrieved case is reused. In the third 'R' stage, the solution is revised to suit the new problem, and in the fourth 'R' stage, the revised solution and the problem are retained for future reuse. Obviously, CBR is naturally suitable for knowledge-based decision making. The success of a CBR system is subject to the suitability of knowledge and the correctness of reasoning.

Reinartz et al. (2001) extended the standard four-stage CBR cycle with two additional stages: *review* and *restore*. In the review stage, the current state of a CBR system and its knowledge containers is monitored and judged; the system, including its knowledge store, is adapted in the restore stage to improve the system performance. Figure 1.3 shows the six-stage CBR cycle.

CBR-based decision support systems can be passive or active. They can be used to aid and support novice or expert-level decision makers, and can be used to help a wide variety of decision-making activities. The major advantage of CBR methods in support decision making is that CBR is considered as a natural reasoning process of human beings. Normally, people are good at using cases but not as good at recalling the right ones. CBR can extend the decision makers' memory by providing the right cases. The major disadvantage of the CBR method is that the solution space cannot be fully explored. As a result, there is no guarantee of an optimal solution for a decision problem.

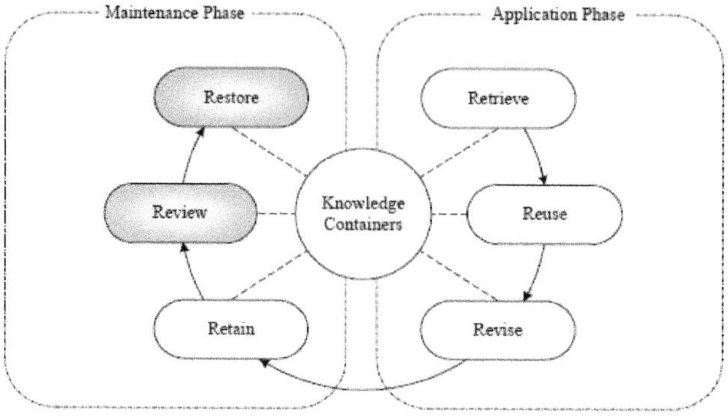

Fig. 1.3 The Six-Step CBR Cycle (Reinartz et al. 2001)

1.5.5 Decision Tree

A decision tree is a graph of decisions and their possible consequences which is used to create a plan to reach the goal of a decision. A decision tree is a predictive model to map observations about an item with conclusions about the item's target value. Each interior node corresponds to a variable; an arc to a child represents a possible value of that variable. A leaf represents the predicted value of the target variable given the values of the variables represented by the path from the root. Figure 1.4 shows an example of a decision tree used to determine the type of a business. It can support strategy making for medium and small businesses in different industries.

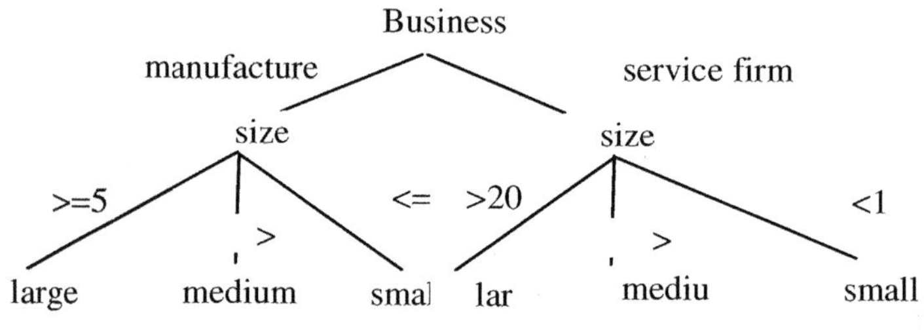

Fig. 1.4 An example of a decision tree

Whatever the decision technique, a critical issue we need to deal with is *uncertainty*. Decision environments and data sources possess various uncertain factors, which result in uncertain relations among decision objectives and entities. Meanwhile, data itself has uncertainty. For example, an individual's preference for alternatives and judgment of criteria is often expressed by linguistic terms, such as 'low' and 'high', which implies a kind of uncertainty. Precise mathematical and inference approaches are not efficient enough to tackle such uncertain variables.

Various uncertain information processing techniques have been therefore developed. Fuzzy decision-making theory, that is, applying fuzzy sets into decision-making activities, is one of these well-developed techniques (Bellman & Zadeh 1970; Kacprzyk et al. 1992; Zimmermann 1987). Research results include fuzzy decision models, fuzzy multi-objective decision making, fuzzy multi-criteria decision making, fuzzy case-based reasoning, fuzzy decision tree, fuzzy data retrieval, fuzzy associate rules for data mining, and their applications.

1.6 What's New in This Book?

First of all, this book is not another addition to the business management textbooks, nor general theories or technologies of decision making. The information presented in the above five sections serves as a preliminary for you to read through the following chapters of this book, and understand its offering.

1.6.1 The Decision Problems Oriented in This Book

This book is mainly concerned with complex ill-structured decision problems, particularly those with uncertainty, time pressures and high personal stakes. New theoretical models and practical technologies will be introduced to help readers to deal with this kind of decision problems.

There are decision situations where decision problems are perfectly structured and can be easily defined. However, more common decision problems in reality are ill-structured. Take business as an example. Today's companies operate in an extremely complex business environment involving many factors interacting with each other. Typical internal business sectors include market research, product R&D, basic engineering, financial management, cost control, and operational efficiency; external business sectors include market, technological, competitive, political/legal, economic, and socio-cultural. It is crucial for today's business managers to keep aware of all these business sectors in order to properly steer the company.

Business decision making happens in a complex environment, and is often imposed with a high degree of complexity, uncertainty, dynamics, high personal stakes and time pressure. Business decision making grows increasingly ill-structured and demanding of experience. Decision making for ill-structured business problems therefore taxes managers' cognitive abilities to a much higher extent than structured ones. Furthermore, business managers need to identify potential threats and opportunities in advance, and respond in a timely fashion, in order to survive in the fierce business competition.

The research and applications in the DSS area have yielded a wide variety of theories and technologies (Sections 1.3 and 1.4) to deal with different kinds of decision problems. Model-based DSS have been well-studied and successfully applied to solve the structured decision problems for decades. Nevertheless, ill-structured decision problems still lack sufficient efficient decision aids from today's DSS. This book will present readers with a set of new models and techniques to explore, investigate and possibly solve ill-structured decision problems occurring in different domains.

1.6.2 New Models and Techniques for Ill-Structured Decision Problems

In this book, a cognition-driven decision process (CDDP) model is discussed, in order to handle ill-structured decision problems. Based on the CDDP model, a series of techniques (methods, algorithms and systems) are introduced for readers to develop decision support systems in their specific application domain.

Compared to the traditional DSS-based decision process model, the major feature of the CDDP model is cognition oriented. In the traditional DSS-based decision process model, decision makers and computers are separated two parts. The design of DSS is mainly driven by technologies. A decision maker in a DSS simply acts as a user of the computer-based system. The CDDP model incorporates managers (decision makers) as the central part of the system as a whole. The managers' cognition is represented as information objects and used for computer-based information processing. Decision making becomes an integral process comprised of the human cognitive process as well as computer-based information processing. The two sorts of processes are both driven by managers' cognition.

Following the CDDP model is a set of techniques including methods, data model and algorithms in order to implement the CDDP model in practical applications. The techniques included in this book are mainly of four types.

(1) Techniques used to represent and process domain knowledge. According to the CDDP model, two kinds of domain knowledge are concerned: domain ontology and management experience.

(2) Techniques used to parse managers' natural language input. Supported by the new techniques, managers can interact with the system using a natural language (English).

(3) Techniques used to construct data warehouse queries. Data warehouse queries are constructed automatically by the system according to the manager's natural language input.

(4) Techniques used to present information. The information most relevant to the manager's decision problem is retrieved from the data warehouse and then presented to the manager. A navigation-knowledge-guided situation presentation method will be discussed in this book.

As an integration of the above techniques, a decision support system, called FACETS, is discussed in details, including its architecture, functions and evaluation. FACETS was designed and developed totally based on the CDDP model, using the above techniques. To demonstrate the application of FACETS, two illustrative case studies, in business and public health domain respectively, are also discussed.

Chapter 2
Business Intelligence

Business intelligence (BI) systems are data-driven DSS, focusing on the manipulation of large volumes of company data in data warehouses. As other types of DSS, such as model-driven DSS, communication-driven DSS, and knowledge-driven DSS (Section 1.5.3), BI systems suffer from lack of cognitive decision support, despite their powerful data analysis functions. This chapter will discuss the basic BI concepts, architecture and some vendors' BI products. The drawbacks of today's BI are also analyzed in details.

2.1 What Is Business Intelligence?

The term business intelligence means different things in different domains. From technical point of view, BI refers to the process of extracting, transforming, managing and analyzing business data, in order to support decision making. This process is mainly based on large data sets, particularly data warehouse, with the mission of disseminating intelligence or knowledge across the whole organization, from strategic level to tactical and operational level. A typical BI process consists of five key stages (CSIRO 2003):

(1) Data Sourcing

A BI system is able to extract data from multiple data sources, representing different business units, such as marketing, production, human resource and finance. The extracted data must be cleaned, transformed and integrated for analysis.

(2) Data Analysis

In this stage, data is converted into information or knowledge through different data analysis techniques, such as reporting, modeling, visualization and data mining. The results of data analysis help managers to have a better understanding of the environment and make better decisions.

L. Niu, J. Lu, and G. Zhang: Cognition-Driven Dec. Supp. for Business Intel., SCI 238, pp. 19–29.
springerlink.com

(3) Situation Awareness

Situation awareness (SA) is a deep understanding of the current decision situation based on the results of data analysis. SA is a key prerequisite for decision making. BI systems should be able to aid decision makers to develop rich SA about their decision situations.

(4) Risk Assessment

Richer SA can help managers to make prediction about the future, identify threats and opportunities, and thus respond correspondingly. Today's businesses are operated in an increasingly complex environment. Business decision making is more likely to suffer risks from the external and internal environment. Thus, risk assessment is an important function of a BI system.

(5) Decision Support

The ultimate goal of BI is to help managers to make decision wisely, based on the current business data.

2.2 The Architecture of a Business Intelligence System

A typical BI system consists of four levels of components and a metadata management module (Codd et al. 1993; Inmon 2002). The generic architecture of traditional BI systems is shown in Figure 2.1. These different components cooperate with each other to facilitate the basic BI functions: extracting data from company operational systems, storing the extracted data in a center data warehouse, and retrieving stored data for various business analysis applications.

- Operational Systems Level

As the data sources of a BI system, business operational systems are mainly online transaction processing (OLTP) systems which support daily business operations. Typical OLTP systems are customer order processing systems, financial systems, and human resource management systems.

- Data Acquisition Level

This level is a data pre-process component including three phases: extracting, transforming, and loading (ETL). A company usually has different OLTP systems producing huge amounts of data. This data is first extracted from OLTP systems by the ETL process and then transformed according to a set of transformation rules. Transformed data is clean, unified, and aggregated and finally loaded into a central data warehouse. ETL is the most fundamental component of a BI system because the data quality of all other components mainly relies on the ETL process. In the design and development of ETL, data quality, system flexibility, and processing speed are the major concerns.

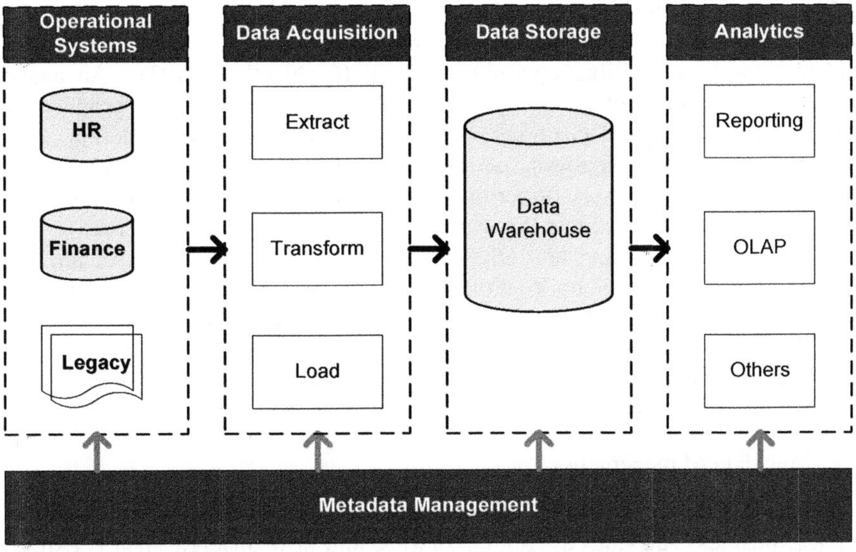

Fig. 2.1 The General Architecture of Current Business Intelligence Systems

- Data Storage Level

The data processed by the ETL component is stored in a data warehouse which is mainly implemented using traditional relational database management systems (RDBMS). A RDBMS is designed to support transaction processing. By contrast, a data warehouse is a subject oriented, time-variant, non-volatile and integrated data store (Inmon 1993)[1]. Data from company OLTP systems are extracted, transformed, and loaded into the data warehouse based on pre-defined schemas. The star schema and snowflake schema are the most popular data warehouse schemas. No matter what kind of schema on which a data warehouse is designed, the data warehouse always includes two basic types of tables: fact tables and dimension tables.

- Analytics Level

Based on the data warehouse, various kinds of analytical applications are developed, which represent the last level: Analytics. BI systems support two basic types of analytical functions: reporting and online analytical processing (OLAP). The reporting function provides managers with different business reports, such as sales reports, product reports and human resource reports. Reports are generated by executing queries into the data warehouse (DW). The DW queries are mainly pre-defined query sentences programmed by the DW developers. Thus, the reports generated by BI systems usually have static formats and contain fixed types of data.

[1] This is the original definition of data warehouses. Today's data warehouse systems can be company wide in scope and the data can be updated over time, for instance, as real-time BI systems.

The most promising BI analytics is OLAP. OLAP allows managers to efficiently browse their business data from different analysis dimensions through slicing, dicing and drilling operations at will (Codd et al. 1993). An analysis dimension is a perspective through which data is presented, e.g., product type, sales location, time and customer. Compared to the reporting function, OLAP supports ad hoc data analysis, i.e. managers have full control of the data by selecting different analysis dimensions of interest to them. OLAP is based on multidimensional data models (known as the snowflake and star schema).

In addition to reporting and OLAP, there are many other types of analytical applications which can be built on the basis of a DW system, such as data mining, executive dashboards, customer relationship management, and business performance management. Technically, these applications are not necessarily built on a data warehouse. However integrating them with DW systems has become a common practice in many practical BI systems.

- Metadata Management

Metadata are special data about other data such as data sources, data warehouse storage, business rules, access authorizations, and how different data is extracted and transformed. Metadata is crucial for producing accurate, consistent information and system maintenance. It affects the entire process of designing, developing, testing, deploying and using BI systems (Caserta 2004; Inmon 2002).

2.3 Analytics of Business Intelligence

The analytics are the core part of a BI system. Evelson and colleagues (2008) summarized BI analytics into eight categories based on a lab-based evaluation of popular BI tools on the market.

(1) Production/operational reporting for pixel-perfect mass report distribution

No matter how much BI self-service end users request, good old-fashioned report development tools, mainly used by professional programmers, remain at the heart of a BI product line. While these tools may also be used to analyze data and produce visual dashboards, they are primarily used for mass distribution of very sophisticated reports like customer statements. Requirements for these products often include pixel-perfect positioning of data and graphics, a scripting language equal in power to a full programming language, and the ability to handle complex headers, footers, nested subtotals, and multiple report bands on a single page.

(2) Ad hoc query tools provide a quick answer to a business question

When report formatting or distribution is not a requirement, and an information management professional just needs a quick answer to a business question like, "How many units of a product were sold yesterday across all stores and outlets?" or, "What were my total sales in 2007 in North America?" simple ad hoc query tools with an intuitive point-and-click user interface (UI) are the way to go.

(3) OLAP tools, when business questions are more about "whys" than "whats"

While reporting and ad hoc query tools are typically used to answer the questions like "What happened?" and"When and where did it happen?", online analytical processing (OLAP) tools are used to answer the questions like "Why did it happen?" and also to perform "What if?" analysis. Otherwise known and "slicing and dicing" analysis (essentially a spreadsheet pivot table on steroids), OLAP tools allow a power user to see any facts (numerical, typically additive numbers, like transaction amounts and account balances) almost instantaneously regrouped, re-aggregated and resorted by any dimension (descriptive elements like time, region, organizational unit, or product line).

(4) Dashboards as an interactive, visual UI — not a reporting or analytical tool by itself

Dashboards should be used as a UI to operational or analytical information. Designed to deliver historical, current, and predictive information typically represented by key performance indicators (KPIs), dashboards use visual cues to focus user attention on important conditions, trends, and exceptions. The term *dashboard* is often used synonymously with *scorecard*, but Forrester defines a scorecard as just one type of a dashboard that links KPIs to goals, objectives, and strategies. Many scorecards follow a certain methodology, such as Balanced Scorecard, Six Sigma, Capability Maturity Models, and others. Other dashboard varieties include business activity monitoring (BAM) dashboards and visualizations of data/text mining operations.

(5) BAM will report on real-time data and process information streams

While a dashboard can be used as a graphical user interface (GUI) component, BAM also captures data and process events (e.g., number of credit applications processed today and number still pending in a queue), correlates and aggregates them into business metrics (e.g., ratios of processed, approved, and rejected applications per hour), and displays the real-time status of the metrics and trailing patterns.

(6) Predictive modeling answers questions about what's likely to happen next

Using various statistical models, these tools attempt to predict the likelihood of attaining certain metrics in the future, given various possible existing and future conditions. One typical predictive modeling class is called market basket analysis, which tries to predict the likelihood of a customer buying a certain product if and when he or she bought another product at a certain store at a certain season, date, and time, given certain economic conditions such as interest rates and price of gas.

(7) BI workspaces enable true end user self-service

While most BI environments attempt to address end user self-service requirements, they still impose many restrictions, such as fixed data models, an inability to add new dimensions on the fly, and sometimes restricted access to production data. Forrester defines a BI workspace as a data exploration environment where a power user can analyze production, clean data with near complete freedom to modify data models, enrich data sets, and run the analysis whenever necessary, without much dependency on IT and production environment restrictions. Some examples of such workspaces are desktop-based multidimensional OLAP (MOLAP) cubes, in-memory data models, or BI software-as-aservice (SaaS).

(8) Guided BI search tools support free form ad hoc queries and analysis

While reporting, ad hoc queries, and OLAP tools work best when one knows the exact business question, they fall short when a user is looking for something that he or she is not quite sure of. A salesperson getting ready for an important client meeting may not know all of the information required to prepare for the meeting and may not be able to effectively construct the appropriate queries to pull the information he/she might need. What works much better is enabling this salesperson to simply enter a few keywords to find relevant customer dimensions in the database, then using a graphical interface to drill into the information he/she wants from a list of possibilities. This effectively solves one of the oldest dilemmas in BI: having to know exactly which questions to ask to get a meaningful answer.

2.4 Commercial Tools

2.4.1 SAS Business Intelligence

Website: *http://www.sas.com/technologies/bi/*

SAS BI offers a full breadth of SAS Analytics capabilities, including statistics, predictive analytics, data and text mining, forecasting, and optimization. These functions are integrated within the business context for better, faster decision making. SAS BI has two components: *Enterprise Business Intelligence* and *Business Visualization.*

The typical functions of SAS BI are as follows.

- Web and desktop reporting

SAS BI supports a wide variety of targeted, fit-to-task interfaces for report building, viewing and distribution for all levels of users across an organization.

- Portal and customizable dashboards

SAS BI provides users an easy-to-use, role-based Web portal, via which users can access aggregated information. It also includes a dashboard development environment, enabling users to create their own dashboards of different styles from virtually any data source.

- Microsoft Office integration

Microsoft Office can be integrated with SAS BI, bringing SAS capabilities in data access, reporting and analytics directly from Microsoft Office.

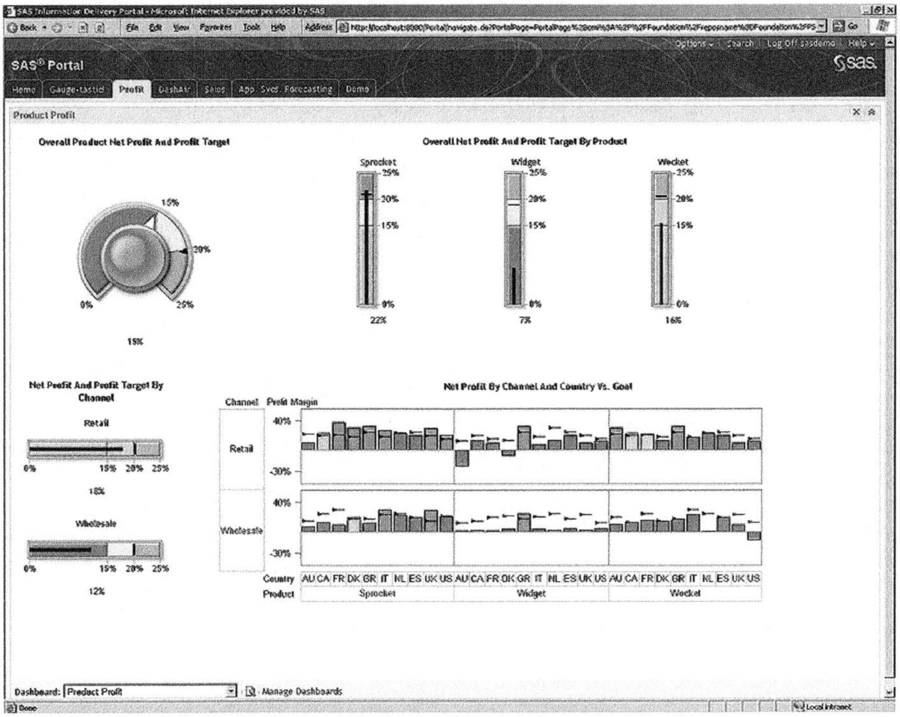

Fig. 2.2 A Screenshot of SAS BI

- Query and Analysis

SAS BI has a user-friendly interface to allow users to easily access and query data on their own without the help of IT staff.

- Interactive business visualization

SAS BI presents data in charts, graphs and geographic maps within multiple BI applications.

- OLAP storage and OLAP data exploration interface

Users can work on their Web browsers to create OLAP cubes and interact with SAS BI to view the multidimensional data from different business dimensions.

- Integrated analytics

Users can access sophisticated analyses directly from their BI interface for decision making.

- Guided analysis.

A dynamic Windows interface can guide users during model development. This function enables business analysts, statisticians and programmers to leverage SAS analytics and efficient processing across all enterprise platforms.

A screenshot of SAS BI is shown in Figure 2.2.

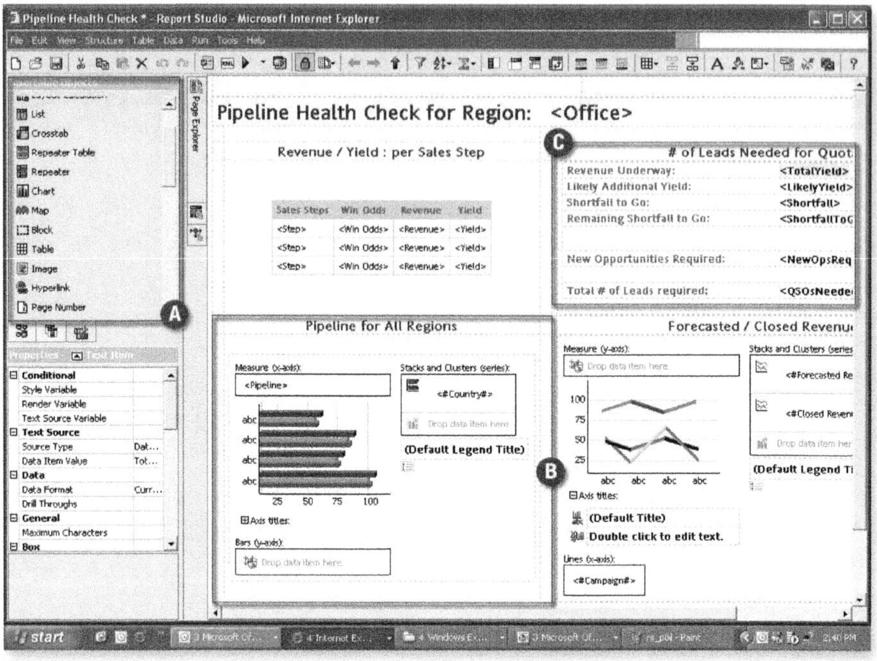

Fig. 2.3 IBM Cognos BI

2.4.2 *IBM Cognos Business Intelligence*

Website: **www.ibm.com**

IBM Cognos BI provides full breath of BI analytics, such as various styles of reporting, analysis, score carding, and dashboards. It has four major components:

- Analysis Studio

In Analysis Studio, users can explore their business data to find trends and comparisons that answer their business questions. The analysis in Cognos is no longer strictly against Cognos Power Cubes but also against relational data sources.

- Report Studio

Report Studio is the platform for power users and professional developers to create formatted reports that contains multiple charts or tabular data sets from multiple subject areas. Additional chart types, such as gauges and maps, that aren't available in Query Studio or Analysis Studio, are also supported in Report Studio. A screenshot of Report Studio is shown in Figure 2.3.

- Query Studio

Query Studio is used for ad hoc report authoring. Users can easily query any data sources (relational, multidimensional or planning data source) to create crosstabs, simple charts or detailed reports. Query Studio also provides formatted templates to give any report a standard corporate layout or logo.

- Metrics Studio

Metrics Studio is used for scorecarding. Users can monitor business performance through different parameters. The idea behind it is to put performance indicators next to the organization's key performance measures (red, orange, green status notation).

2.4.3 SAP BusinessObjects Business Intelligence

Website: www.sap.com

SAP BusinessObjects provides a full spectrum of BI functionality, ranging range from reporting, query, analysis, dashboards and visualization, to intuitive discovery and advanced predictive analytics capabilities, as well as data quality and extract, transform, and load functionality. The components and functions of SAP BusinessObjects are categorized as follows.

- Information Infrastructure

The information infrastructure of SAP BusinessObjects allows IT department to extend BI to any application or process in any environment. It provides following functions: Auditing, BI content search, BI widgets Central management console, Encyclopedia, InfoView, Java portal integration kit, Life-cycle management, Publishing, Process tracker, Query as a Web service, Microsoft Office SharePoint portal integration kit, Software development kits, and Universe designer.

- Reporting

SAP BusinessObjects contains a reporting tool: Crystal Reports providing users an interface to connect to virtually any data source, design and format interactive reports, and share them internally and externally. The final reports can be delivered via SAP BusinessObjects Enterprise, Crystal Reports Viewer, and Microsoft Office documents.

- Query, Reporting, and Analysis

There are two tools in SAP BusinessObjects, Desktop Intelligence and Web Intelligence, allowing users to perform ad hoc query, reporting and analysis, without having to understand complex database languages and underlying structures. Figure 2.4 is a screenshot of SAP BusinessObjects Web Intelligence.

- Dashboards and Visualization

SAP BusinessObjects has nine software tools to support information visualization, such as Xcelsius Enterprise, Dashboard Builder, and VizServer. This set of tools also provides professional developers a software development kit to customize advanced visualization interfaces.

- Advanced Analytics

SAP BusinessObjects Voyager is the OLAP interface through which users can explore multidimensional data. SAP BusinessObjects Predictive Workbench is a software tool enabling users to conduct predictive analysis. SAP BusinessObjects Set Analysis is a software tool enabling users to conduct clustering analysis.

- Search and Navigation

With SAP BusinessObjects, business users can search BI contents from internal and external resources, such as structured databases, business intelligence (BI) systems, unstructured company and text content, search engines, and the Web.

2.5 Limitations

Business intelligence is promising to turn 'data' into 'knowledge' and help managers survive data tsunami and eventually succeed in decision making. However, BI systems are essentially data-driven DSS. Current BI systems can only partially support managers' work (Singh et al. 2002). The emphasis of BI analytics is manipulation of large volumes of business data, rather than supporting managers' decision making from the cognitive perspective.

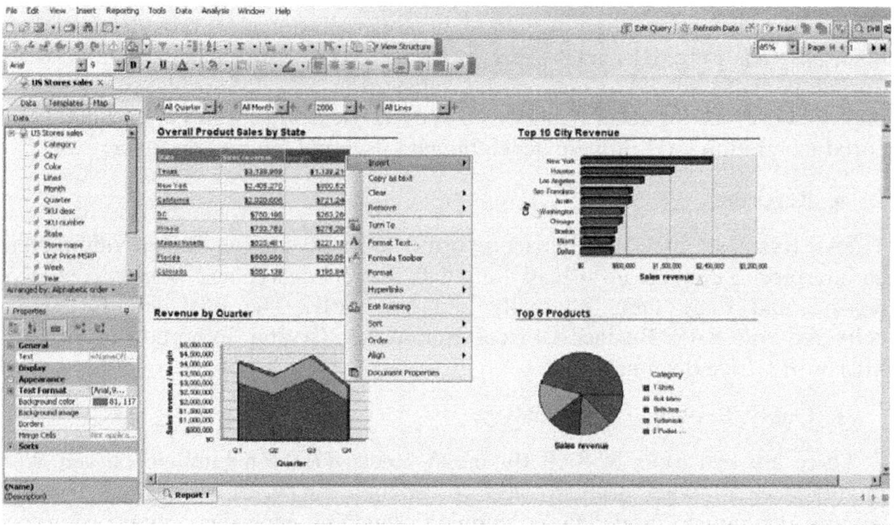

Fig. 2.4 SAP BusinessObjects Web Intelligence

A BI system is capable of providing managers with huge amounts of internal and external business data, but more data does not equal more valuable information (Endsley et al. 2003). On one hand, the reporting function is mainly pre-defined information representation. That is, business reports are generated in fixed types and formats by executing pre-defined queries into the data warehouse. The pre-defined reports are efficient and effective in reporting repetitive and structured business events, for example periodical (daily, weekly, monthly, quarterly and annual) product sales. However, pre-defined reports are not as flexible as many other ill-structured events require, for example, unpredictable marketing campaigns. On the other hand, OLAP-based ad hoc analysis gives managers the full control of their data. Managers can easily obtain any data of their interest by selecting analysis objects and customizing analysis dimensions. Nevertheless, mangers often feel lost when confronted with a large body of business data concerning a decision problem (Resnick 2003).

A recent survey by Economist Intelligence Unit (2006) shows that 73 per cent of senior managers agreed that it is important to have less but more timely data to improve the quality and speed of decision making. This result corresponds to the research result by Sutcliffe and Weber (2003) about knowledge accuracy. Their research implies that having a lot of facts about a decision situation is less important than having a clear and consistent overview. Resnick (2003) criticizes the current executive dashboard design (a type of BI application) for emphasizing improvement on data analysis functionality while falling short of cognitive engineering consideration. More recently, an industrial report from InfoWorld Media Group shows that 'BI has a reputation for being a resource sink that delivers reports almost no one reads. It doesn't have to be that way. And you can no longer afford to let it be' (Gruman 2007, p. 22).

2.6 Summary

The identification of the advantages and disadvantages of traditional BI system motivated us in this research to seek a new way to better support managers' work, particularly for handling ill-structured decision problems. This book aims to provide models and techniques to facilitate cognitive decision support on the basis of BI platforms. Thus, it is necessary to look at the nature of cognitive decision support from the cognitive psychology perspective. In next chapter, relevant concepts and decision models in cognitive psychology will be reviewed.

Chapter 3
Managerial Cognition

In naturalistic decision making, successful decisions are mainly based on the sufficient situation awareness (SA) and mental models of decision makers. With rich SA and mental models, a decision maker is able to understand the current decision situation, identify potential threats and opportunities, and predict the future. In this chapter, we will introduce the basic concepts and models of SA, mental models and naturalistic decision making, and the implications of them to investigating the problems of the decision making model of classical DSS.

3.1 The Concept of Cognition

In cognitive psychology, cognition refers to cognitive (mental) processes (functions) involving acquisition, maintenance and usage of knowledge. A cognitive process is considered as a process of human information processing '… by which you [an individual] take information, pick it over, play with it, analyze it, put it together, reorganize it, judge and reason it, make conclusions, plans and decisions, and take action' (Jaques 1996, p. 18). Cognition can take a number of forms such as perception, attention, pattern recognition, learning, memory, language processing, problem solving, thinking and reasoning (Baum 2004; Lycan & Prinz 2008; Sternberg 2006).

The resultant product of cognition is knowledge, which can be of different types, such as beliefs, mental models and SA. Knowledge is stored and maintained in memory which is either long-term memory (lasting from days to a lifetime) or working memory (lasting around 20 seconds) (Bahrick et al. 1975). Knowledge is typically utilized by people to make decisions and judgments (Plous 1993). In the naturalistic setting, people are capable of making decisions solely based on the knowledge residing within their minds. How experts (proficient decision makers) make decisions in real-world contexts that are meaningful and familiar to them is the major concern of naturalistic decision making (NDM) research (Lipshitz et al. 2001). People's abilities to make naturalistic decisions mainly rely on two kinds of knowledge: SA and mental models. In this section, we will briefly review the existing work about SA, mental models and NDM.

L. Niu, J. Lu, and G. Zhang: Cognition-Driven Dec. Supp. for Business Intel., SCI 238, pp. 31–37.
springerlink.com

3.2 Situation Awareness

The concept of SA was initiated in the military aircraft domain and extended to air traffic control, nuclear power plants, and other tactical and strategic systems (Endsley 1995b). In aviation, SA mainly refers to the pilot's knowledge about the aircraft itself and its environment (Emerson et al. 1987; Hamilton 1987; Vidulich 1995). Sarter and Woods (Sarter & Woods 1991) describe SA as 'the accessibility of a comprehensive and coherent situation representation which is continuously being updated in accordance with the results of recurrent situation assessments' (p. 52).

Endsley (1995b) proposed a generic model of SA in terms of information processing (Figure 3.1). She suggested that SA can be divided into three levels or steps of mental representation.

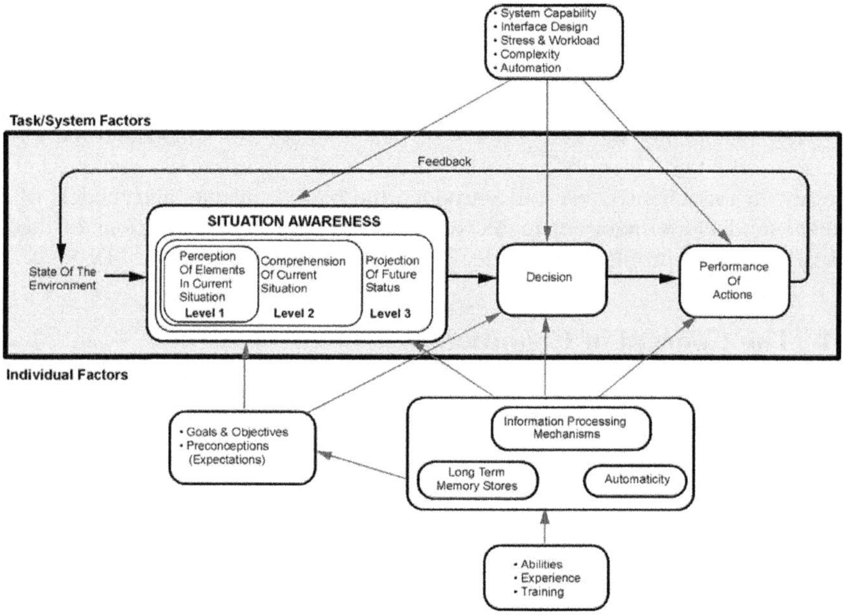

Fig. 3.1 Endsley's Situation Awareness Model (Endsley 1995b)

Level 1: SA is the decision maker's perception of the status, attributes, and dynamics of relevant elements in the environment (the decision situation). Level 1 SA is the lowest and most basic level of SA. Achieving Level SA involves basic information detection processes.

Level 2: SA is the decision maker's comprehension of the perceived information, i.e., Level 1 SA. Level 2 SA is achieved through pattern recognition, interpretation, and evaluation. Level 2 SA results a comprehensive picture of the environment.

The highest level of SA is the decision maker's projection: Level 3 SA. Level 3 SA is the decision maker's ability to predict the future status of the environment.

Endsley's SA model also shows various internal and external factors affecting the process wherein SA is developed. The development process of SA is referred to as *situation assessment*. In the SA model, situation assessment is an information processing process within the decision maker's mind. This process can be enhanced by means of appropriate technologies. For instance, a case study by Endsley and co-researchers (2003) demonstrates that different user interface designs result in different degrees of situation assessment in aviation control. In this research, we proposed the situation retrieval theory and developed relevant IS techniques to assist managers to develop SA according to their decision situations.

Situation awareness is believed to be an essential prerequisite for people's decision making in any complex and dynamic situation (Endsley 1995a; Flach 1995; Sarter & Woods 1991; Smith & Hancock 1995). A close relationship exists between SA and decision making: richer SA is more likely to lead to good decisions and then to good performance (Stanners & French 2005).

Simply put, SA is about knowing what is going on around the decision maker. In this book, we particularize the concept of SA in business domain in that SA is a state of knowledge reflecting the manager's understanding of the current decision situation. This understanding may include sensing and comprehending different business information, such as sales data, product information, emerged competitors, and government policies. It may also reflect the manager's prediction about the potential threats, opportunities, and possible solutions. SA creates a big picture of the company within the manager's mind and enables the manager to be capable of predicting the future and of making decisions.

3.3 Mental Models

Mental models are 'psychological representations of real, hypothetical, or imaginary situations' (Johnson-Laird et al. 1998). Mental models are commonly referred to as deeply held assumptions and beliefs that enable individuals to make inferences and predictions (Chen & Ge 2006; Chen & Lee 2003; Johnson-Laird et al.1998). Rouse and Morris (1985, p. 351) defined mental models as 'mechanisms whereby humans are able to generate descriptions of system purpose and form, explanations of system functioning and observed system states, and predictions of future states'.

A mental model is useful in that it provides (1) a mechanism for guiding attention to relevant aspects of the situation, (2) a means of integrating information perceived to form an understanding of its meaning and (3) a mechanism for projecting future states of the system based on its current state and an understanding of its dynamics (Endsley et al. 2003).

Mental models and SA are different in their points of reference (Sarter & Woods 1991). Mental models use finite number of elements and algorithms to represent systems or devices, whereas SA is a dynamic representation of open systems. Mental models reflect people's past experience which act as the basis and guidance for adequate situation assessments (Endsley 1995a; Sarter & Woods 1991). People need both rich SA and mental models to understand the decision situation, to anticipate the near future status of the environment, and then to succeed in decision processes.

Mental models are a type of tacit knowledge which can be elicited from people's minds using cognitive mapping (Ackermann et al. 1992). Cognitive mapping is a technique used to structure accounts of problems. Cognitive mapping produces a pictorial representation of the user's problem: cognitive maps (Eden 1988). Users can organize their concepts (ideas), and the interrelationships between concepts in a cognitive map. Figure 3.2 is an example of cognitive map showing how the notebook sales are affected by different factors. Each node denotes a concept and each directed line denotes a causal relationship between two concepts. For instance, customer service and advertisements affect market share; market share affects notebook sales.

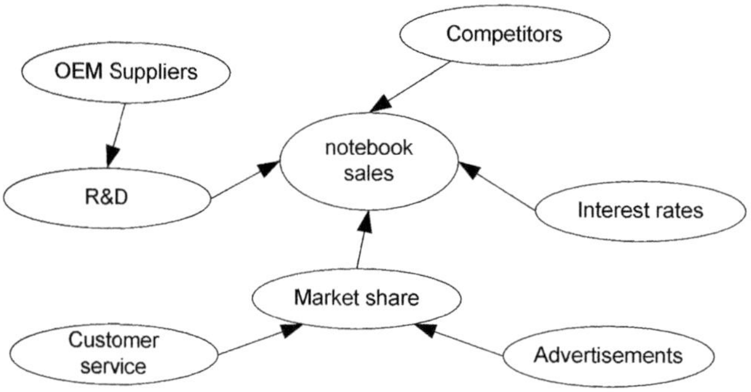

Fig. 3.2 An Example of Cognitive Map

Cognitive maps are also variously referred to as mental maps, concept maps, schemata, mental models or cause maps. Cognitive mapping has been studied and applied in different fields: operation research (Carlsson & Fuller 1996; Langfield-Smith & Wirth 1992), software engineering (Siau & Tan 2005), accounting (Lee & Kim 1997), and strategic management planning (Carlsson & Fuller 1996).

3.4 Naturalistic Decision Making

In the study of decision making, NDM has been receiving more interest recently along with other theories such as CDM, BDT, JDM, and ODM. NDM focuses on investigating how proficient decision makers make decisions in their familiar decision situations (Lipshitz et al. 2001). NDM is a descriptive decision theory. At the other end of the spectrum is normative decision theory, e.g., CDM. In CDM, decision makers are considered as rational human and the decision making is a 'choice' process. Compared to CDM, NDM is based on a 'matching model' and the decision maker has only bounded rationality. NDM is shaped by five essentials (Lipshitz et al. 2001).

(1) Proficient Decision Makers

NDM is concerned with the decision processes of proficient decision makers in their familiar decision situations. Proficient decision makers are people who have rich relevant knowledge or experience regarding decision making. Decision makers' experience is domain specific and forms the basis of their naturalistic decision making.

(2) Situation-Action Matching Decision Rules

Proficient decision makers make decisions via matching processes, instead of choice processes in CDM. When presented with a decision situation, several options will automatically emerge in the decision maker's mind based on his/her past experience. The decision maker then quickly screens most of the options by comparing them against a standard, rather than with one another. An option is selected or rejected based on its compatibility with the situation. The screening process relies much more on pattern matching and informal reasoning than on analytical reasoning.

(3) Context-Bound Informal Modeling

The decision models of NDM tend to be informal and context-specific, that is, closely related to a specific application domain. This is because NDM is mainly based on decision makers' domain-specific experience.

(4) Process Orientation

In CDM, the major problem is how to optimize the output according to a specified input. Rather than the input and output of decision making, NDM is more concerned with the process, particularly with the information decision makers actually search, understand, and use during decision process.

(5) Empirical-Based Prescription

In normative decision models, solutions can be prescribed during decision processes irrespective of the intended recipient's actual ability to perform them, i.e. 'ought' can be divorced from 'is'. In NDM, 'ought' cannot be divorced from 'is': prescriptions are useless if they cannot be implemented, although they are optimal in some formal sense. Thus, NDM researchers believe it is not necessary for prescriptions to be optimal, as long as they are good enough for the current decision situation. NDM prescriptions are derived from descriptive models of domain expert behavior, which are more feasible than the optimal ones from choice models.

Among a number of NDM models, the recognition-primed decision (RPD) model is the prototypical one (Lipshitz et al. 2001). The RPD was developed by Klein and co-researchers (1989) during a study of firefighters, which is shown in Figure 3.3.

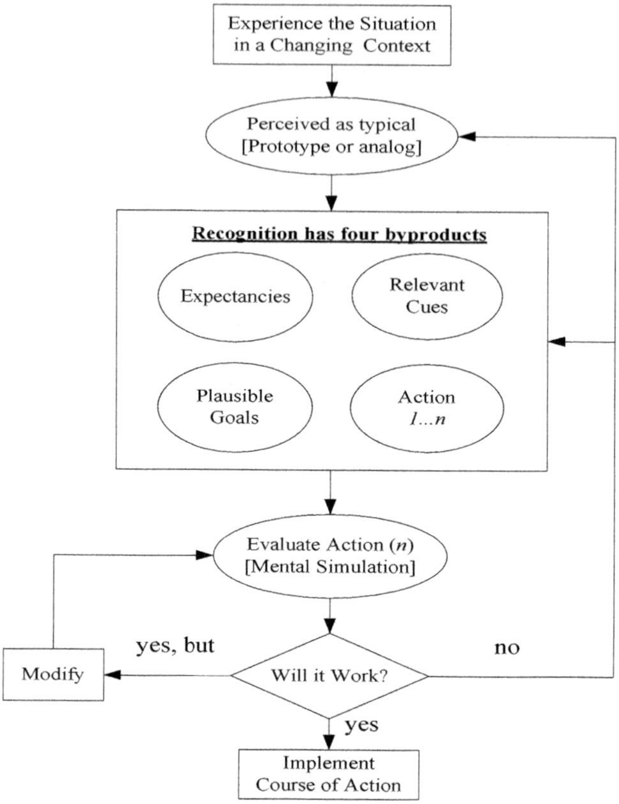

Fig. 3.3 RPD Model (Klein et al. 1989)

The decision process under the RPD model is based on decision makers' SA and mental models. When presented with a new decision situation, the decision maker will intuitively recognize the current situation through developing concurrent SA about it. SA provides the decision maker with an integrated overall picture of the current situation. This picture is a 'pattern' extracted from the current situation based on 'SA' and it is 'matched' with pre-held 'mental models' through pattern recognition. The matched mental model leads directly to a potential solution to the current situation. The feasibility of the potential solution is mentally evaluated against the current decision situation. If the decision maker believes the potential solution will also work in the current decision situation, the potential solution is then implemented. Otherwise, the decision maker will modify or discard the potential solution according to the degree in which it is feasible to the current decision situation. In the RPD model, the processes of situation recognition and pattern matching reply on the decision maker's SA and mental models. Rich and reliable SA and mental models are direct driving forces for decision making.

As two different decision-making theories, CDM and NDM have different applicability for different decision situations. CDM has been successfully applied to solve structured decision problems with well-defined goals, conflict resolution, computational complexity, and requiring optimization and justification. For unstructured or semi-structured decision problems, particularly those with time pressure, uncertainty and ambiguity, NDM is more applicable if experienced decision makers are available.

3.5 Summary

A decision maker's SA and mental models are two prerequisites for successful decision making. Thus, to develop relevant models and technologies to enrich decision makers' SA and mental models is supposed to be a key consideration of designing contemporary DSS. Rich SA and mental models are not the guarantee of successful decision making. However, the strong relationships between SA and mental models and decision making indicates that the better SA and mental models are more likely lead to better decisions.

Chapter 4
Cognition in Business Decision Support Systems

Cognition plays a key role in decision making for complex, ill-structured situations. Cognitive decision support is one of the major objectives in the design and development of DSS. In Chapter 3, we analyzed some important concepts and models in managerial cognition. Managers' rich SA and mental models are two prerequisites for and likely lead to successful decisions according NDM theory. In this chapter, we will look at the role of managers' cognition on decision-making processes from an information system (IS) perspective. The first two sections of this chapter discuss the basic characteristics of decision-making tasks in business domain, and how managers' cognition can influence on business decision making. The last section summarizes some typical systems that take cognition as an important consideration.

4.1 Complex Nature of Business Decision Making

Anthony (1965) categorized management activities into three levels: operational control, management (tactical) control, and strategic planning (Section 1.2). Roughly speaking, management activities of operational control, management control and strategic planning respectively correspond to structured, semi-structured and unstructured problems. However, due to the complexity of business management, there are no definite boundaries between different management activities. For instance, a specific strategic planning activity might be essentially of unstructured nature, but with a proportion of structured characteristics.

Structured management problems are relatively easy to define, and model-based DSS have been well-studied and successfully applied to solve this kind of problems for decades (Gordon & Pinches 1984; Jensen & Bard. 2003). However, ill-structured (semi-structured and unstructured) management problems, particularly strategic planning, still lack sufficient efficient decision aids from today's DSS. Singh and co-researchers (2002) found that the support scope of DSS is surprisingly narrow, with most of the information serving well-defined needs and drawing from internal databases. Much less is done to support the parts of strategic planning that are less structured or require data that are external or non-machine resident. Their research shows that no more than 40 per cent of the strategic management process is supported or partially supported by executive support systems.

L. Niu, J. Lu, and G. Zhang: Cognition-Driven Dec. Supp. for Business Intel., SCI 238, pp. 39–50.
springerlink.com

The lack of decision support of DSS to ill-structured decision problems can be attributed to the complexity of managers' roles. Managers' specific roles include identifying problems and opportunities, working out appropriate solutions to react to problems and opportunities, developing business goals and strategies, establishing social networks, developing and maintaining relationships with different partners (Hoven 1996; Welter 1988). Managers' roles in organizations were formally modeled by Jaques's (1976) stratified systems theory. The stratified systems theory describes a bureaucratic system at eight strata levels corresponding to eight managerial roles respectively. Each organizational stratum is shaped by task uncertainty and time-span of discretion: the targeted completion time of the longest task or task sequence in a managerial role. Jaques found that the time span increases as one moves up the organization, such as 1 day to 3 months (Level 1), 3 months to 1 year (Level 2), 1 year to 2 years (Level 3), 2 years to 5 years (Level 4), 5 years to 10 years (Level 5) and so on. Level 1 tasks are mainly concrete tasks conducted by front line managers and they have minimal uncertainties, compared with Stratum 4 tasks such as corporation goal formulation and strategic planning. With stratum level increases, task uncertainties also increase. Managers employ diverse techniques to deal with uncertainties, such as trying out different ways of carrying out tasks (Level 1), cognitively simulating tasks in the mind (Level 2), using forecasting systems (Level 3), constructing portfolios of departments and systems (Level 4), scenario analysis (Level 5) and so on. Jaques points out tasks become more complex due to increased uncertainties towards higher level managerial strata, and the increasing task complexity demands increasingly complex cognitive processes.

Regarding the complexity of ill-structured decision problems, Mintzberg and co-researchers (1976) described strategic planning as novelty, complexity and open-endedness. They refer to strategic decision making as a 'groping process'. At the beginning of this process, managers usually have very little understanding of the decision situation with which they are confronted, and they only have a vague idea about the potential solution. Before a final decision is made, managers have to grope through 'a recursive, discontinuous process involving many difficult steps and a host of dynamic factors over a considerable period of time' (Mintzberg et al. 1976).

This is not the decision making under uncertainty of the textbook, where alternatives are given even if their consequences are not, but decision-making under ambiguity, where almost nothing is given or easily determined (Mintzberg et al. 1976, p. 136).

Today's companies operate in a turbulent business environment where different sectors interact with and affect each other. Walters and co-researchers (2003) summarize in six internal business environment sectors (market research, product R&D, basic engineering, financial management, cost control, and operational efficiency) and six external ones (market, technological, competitive, political/legal, economic, and socio-cultural). For the survival of a company, the executive needs to keep abreast of each sector of the environment. Moreover, the speed and quality with which business decisions must be made has increased substantially with the trend of the globalization of economy (Ditsa 2003; Resnick 2003).

The above discussion shows that today's business decision making happens in a complex environment, and is often imposed with a high degree of complexity, uncertainty, dynamics, high personal stakes and time pressure. As Courtney argues '...organizational decision environments have always been complex and ill-structured, the environments of the near future will be even more so' (Courtney 2001, p. 18), business decision making grows increasingly ill-structured and demanding of experience. Decision making for ill-structured business problems therefore taxes managers' cognitive abilities to a much higher extent than structured ones. Only with rich situation awareness and solid mental models about their decision situations can managers have good opportunity to be successful in managing their business.

4.2 Cognition in Business Decision Making

Managers' cognition plays an important role in decision making, as has been noted by many researchers. In behavioral organization theory, managers' cognition acts as a filter between inter-organizational and intra-organizational environments, which helps managers to search for selective information concerning functions of the business and certain organizational actions (March & Simon 1963). Similarly, based on a survey of 12 Fortune 500 companies, Donaldson & Lorsch (1983) concluded that senior executives simplify business reality by employing interrelated beliefs to filter irrelevant information. The simplified business environment helps executives to gain better understanding of their companies during strategic decision making. This conclusion was confirmed by Porac & Thomas (1990).

Mintzberg (1973) categorized managers' work into ten different roles and connects them with managers' mental models. He found that managers spend most of their time communicating with other people and thinking, by which a series of mental models are built. Managers' mental models are related to different issues, decision situations, problems and company internal and external environments. During the decision process, managers tend to use their past mental models to comprehend the current problem and to test alternative solutions.

Isenberg (1984) observed that higher level decision making is mainly based on managers' intuition rather than 'choosing' the best one from a number of identified alternatives. Managers are skillful at using historical experience to envision future scenarios of the company, by which they predict potential threats and possible opportunities. In dynamic, ill-structured environments, managers have little time to conduct thorough rational reasoning. Ironically, managers tend to quickly assess decision situation by comparing a current decision problem with their past decision scenarios, which leads to a quick, satisfying decision (Schmitt 1997).

Managers' cognitive abilities are nevertheless subject to many cognitive biases. A cognitive bias is a distortion pattern in the human mind which leads to a perception, judgment, or reliability that deviates from the reality (Pohl 2004). Cognitive biases might be useful in certain circumstances, but they are more likely to cause serious mistakes in decision making. For example, people tend to accept new information that confirms their preconceptions and avoid conflicting ones. Senge (1990) found that many good business plans failed to be carried out simply because of their incompatibilities with managers' mental models. When people retrieve past experience, they have a tendency to overestimate recent events because these events are relatively easier to recall from memory. People also have a bias toward thinking that the more information they can obtain, the better decision they can make. Russo and Schoemaker (1990) described ten most common mistakes in decision making related to cognitive biases: plunging in, frame blindness, lack of frame control, overconfidence in your judgment, shortsighted shortcuts, shooting from the hip, group failure, fooling yourself about feedback, not keeping track and failure to audit your decision process. Most cognitive biases are hard to avoid and they are attributed to different psychological biases. For example, judgmental biases are caused by judgmental rules and heuristics employed by people to reduce difficult mental tasks to simpler one (James H. Barnes 1984). However, well-designed information systems are helpful for people to overcome some bad cognitive biases (Chen & Lee 2003). In this book, a navigation-knowledge-guided situation presentation method will be discussed to help managers easily recall and reuse past experience and to provide support to their thinking process.

4.3 Cognition Oriented Information Systems

Theoretically and practically, the existing literature reviewed in previous sections of this chapter advocates a strong argument that today's DSS should be designed to support managers' work from a cognitive perspective, in order to assist decision makers to overcome the disadvantage of cognitive biases, reuse past experience, enhance mental models, develop SA, and deal with ill-structured decision problems. In this section, we review existing work about specific systems, related models and concepts for cognitive decision support in DSS. As a wider appreciation of previous work in this research, we also briefly review other areas related to supporting or utilizing human cognition in a broad sense. These areas include case based reasoning (Section 1.4.4) and natural language interface to databases (NLIDB).

4.3.1 Cognitive Decision Support Systems

Cognitive maps, as a knowledge representation technique of human mental models, have received wide research attention in DSS community. A number of DSS have been developed in past research projects to support manipulation of cognitive maps.

An early DSS called SPRINT (Strategic Plan and Resource Integration), was developed to support strategic management by Carlson and Ram (1990). SPRINT can be used by managers to explicitly represent planning models which are of implicit nature in managers' minds. The visual representation of planning models is based on manages' mental models. The concept nodes and links between concepts can be created by managers according to their understanding and thoughts about their decision problems. SPRINT also supports heuristic rules and goal-oriented communication between different managers. The cognitive aspects of SPRINT lie in supporting the visual representation and dynamic creation of managers' mental models regarding business plan formulation. Although managers' cognition is supported in terms of information systems to a limited degree, SPRINT represents one of the early research efforts toward cognitive decision support in the DSS community.

A conceptual DSS called *Cognitive Lens Support System* was described by Yadav and Khazanchi (1992). They proposed the concept *cognitive lens* as the description of mental models from an IS perspective. A cognitive lens acts a filter to convert information into a set of constructs and their interrelationships of the real world. The cognitive lens support system revolves around inquiry of cognitive lenses stored in a database. They proposed three categories of IS function for inquiry of cognitive lenses: introspective, dialectic, and eclectic. The introspective function allows managers to examine their past experience for a specific decision problem. The dialectic function allows managers to compare their own experience with others. The eclectic function allows managers to aggregate multiple pieces of experience. The major argument for the cognitive lens support system is that the IS functions developed based on managers' cognitive orientation will facilitate better understanding of ill-structured problems. Compared to previous research, the cognitive lens support system illustrates a more comprehensive analysis about the significance and IS techniques of supporting managers' thinking process for business decision making, although at a conceptual level and lacking empirical validation.

Following the cognitive lens support system (Yadav & Khazanchi 1992), Chen and Lee (2003) developed a cognitive DSS for strategic decision making. Similarly, their system also includes three supporting modules: retrospective, introspective, and prospective. The retrospective module provides managers with tools to manage business cases, experience, other people's views, speculations, and even rumors. The introspective module is used to explore and represent the managers' mental models. The prospective module provides managers with aids in forward thinking by creating and managing future business scenarios. An exploratory assessment was conducted to evaluate this system by interviewing real business executives from three different industrial sectors. The evaluation results show that cognitive decision support can be effectively delivered by aiding managers in explicitly representing and exploring their mental models.

Cognitive maps, particularly fuzzy cognitive maps, are also employed as a reasoning mechanism for developing domain-specific DSS. Lee and Kim (1997) developed a bidirectional (downward or upward) inference system based on fuzzy cognitive maps to solve highly unstructured problems in stock investment domain. Noh and co-researchers (2000) combined cognitive map technique with case-based reasoning to solve credit analysis problems. Konar and Chakraborty (2005) proposed a unsupervised learning and reasoning model based on fuzzy cognitive maps which is implemented with Petri nets. More recently, Stylios and co-researchers (2008) used fuzzy cognitive map technique to assist medical professionals in crucial clinical judgments.

Recognizing the implication of managerial intuition in handling dynamic, ill-structured business problems, Kuo (1998) proposed an ecological cognitive model of managerial intuition for executive support system (ESS) development. This model enables managers to intuitively assess the situation through perception-action cycles. During the perception-action cycles, sensorimotor (perception and actions) is combined with the memory processors (mental models). Cognitive decision support of ESS is reflected in modeling the ecology of managers, i.e. the interplay between human and environment. The claimed contribution of the ecological cognitive model is this model can be used to guide the development of practical ESS with consideration of the managerial intuition.

The impact of computerized cognitive aids was tested in the context of strategy execution process by Singh (1998). This research identified two sorts of specific cognitive requirements in strategy execution processes: memory support and strategy support. The former is intended for compensation for managers' limited attention resources. The latter is for managers' monitoring abilities. Positive and significant relationships were found between the efficiency and effectiveness of the strategy execution process and computerized cognitive aids.

This research has a broader concern of cognition in addition to mental models. For example, the manager's SA is represented as a set of natural language sentences, is analyzed using natural language processing techniques, and is utilized to retrieve experience. Furthermore, mental models are represented and used to formulate information needs. The cognitive processes, such as situation assessment, thinking and decision formulation are also supported by the system.

4.3.2 Case-Based Reasoning Systems

Case-based reasoning (CBR) has been introduced in Section 1.4.4. CBR is based on manipulation of cases, as the representation of historical experience of problem solving (Reisbeck & Schank 1989). Case reuse is the key topic in CBR. A general case reuse process consists of case retrieval, adaptation, reuse for solving a new problem, and retention as a new case (Aamodt & Plaza 1994).

Cases are a kind of knowledge acquired by field experts during problem solving processes (Bergmann 2002). Problem solving is a typical cognitive process of a decision maker. Hence, by their nature, cases can be considered as human knowledge originating from decision makers' mental constructs. In this sense, cognitive decision support is also reflected in CBR systems. Our research benefits from the CBR technique in that they are based on the similar idea: knowledge reuse, i.e., using past knowledge to handle new situations. However, the differences between this research and CBR are also evident. Instead of cases, we focus on other types of knowledge: mental models, SA and ontology. We also focus on decision support rather than solution generation.

4.3.3 Natural Language Interfaces to Database

Another related area to this research is natural language interface to database (NLIDB). A NLIDB system is an information system that allows users to retrieve information from a database through inputting queries in the form of a natural language, for example, English (Androutsopoulos et al. 1995). The research of NLIDB started with the progress of natural language processing (NLP) and it now has become one of the most successful applications of NLP. According to the mechanism of query construction, NLIDB systems can be classified into three basic types: pattern-matching systems, syntax-based systems and semantic grammar systems (Androutsopoulos et al. 1995).

4.3.3.1 Pattern-Matching NLIDB Systems

A pattern-matching NLIDB system applies a set of hard-wired rules to a natural language input (question) and then directly formulates a database query (Androutsopoulos et al. 1995). Suppose a table called *T_Partners* in a database contains contact information of business partners such as Company Name, President, and Telephone Number. A rule used in the NLIDB system might be as follows.

A pattern:

"President/CEO" of <A Company Name>"

A query corresponding to the above pattern:

SELECT *President* FROM *T_Partnter* WHERE *Company_Name* = *<A Company Name>*

If a user inputs into the system "*What's the CEO of ABC Ltd.?*", then the system will generate a query:

SELECT *President* FROM *T_Partnter* WHERE *Company_Name* = '*ABC* Ltd.'

The results retrieved from the database by executing the above query will be returned to the user.

4.3.3.2 Syntax-Based NLIDB Systems

A syntax-based NLIDB system employs a syntactic parser to identify the constituent tree of the question sentence input by a user (Androutsopoulos et al. 1995). The constituent tree is then mapped to a database query expression based on pre-defined syntactic grammars.

A syntactic parser is a software tool which can analyze natural language sentences syntactically. Parsers are built on the basis of a set of grammar rules and dictionaries. Parsers are generally language-specific due to the characteristics of the grammar rules and the dictionaries. The output of a parser for a sentence is the syntactic representation of that sentence. The syntactic representation mainly contains three sorts of information: part of speech (Voss & Post 1988) of each word, phrases and relationships between words or phrases (Roth 2004).

The part of speech of a word in a sentence is its linguistic category, which explains how a word is used in a sentence. There are eight parts of speech defined in traditional English grammar: the *verb*, the *noun*, the *pronoun*, the *adjective*, the *adverb*, the *preposition*, the *conjunction*, and the *interjection* (MacFadyen 2007).

Example 4.1. A sentence

Fat cats like fresh fish.

With parts of speech tags:

Fat (adjective) *cats* (noun) *like* (verb) *fresh* (adjective) *fish* (noun).

Words form phrases. A *phrase* is a group of two or more words which are grammatically linked together without a subject and predicate (MacFadyen 2007). Most parsers can also extract phrases. This process is called *chunking*. As POS tags for words, phrases are also categorized according to their chunk tags (syntactic roles) such as noun *phrase* (NP), *verb phrase* (VP), *prepositional phrase* (PP) and *adverb phrase* (ADVP). For the sake of simplicity, we use term POS to refer to both chunks and parts of speech discussed previously. For a complete description of POS tags and chunk tags, please refer to Penn Treebank project website[2].

Parsers also generate constituent representations for sentences being analyzed. The constituent representation of a sentence is a tree structure called constituent tree. Relationships between words or phrases are represented in the constituent tree. The constituent tree of Example 4.1 is shown in Figure 4.1.

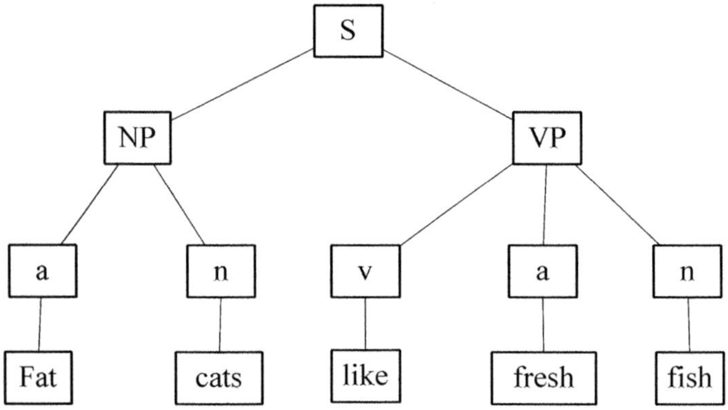

Fig. 4.1 Constituent Tree Example

Syntactic tags used in Figure 4.1 are S: sentence, NP: *noun phrase*, VP: *verb phrase*, a: *adjective*, n: *noun* and v: *verb*.

[2] Information about Penn Treebank Project is accessible at http://www.cis.upenn.edu/~treebank/

Based on the constituent tree, parsers use a set of grammar rules to generate the syntactic relationships between words or phrases. For instance, in Figure 4.1, the parser can find a relationship: *cats* is the subject of *like* (the verb).

Natural language parsing is a support technique for many other applications such as question-answering, machine translation, information extraction and spelling correction. However, natural language parsing is also a challenging research topic in natural language processing due to the ambiguity and variation characteristics of natural language. The research and development of syntactic parsers has received increasingly wide attention both in academics and in practitioners. Many syntactic parsers have been developed for research or commercial purpose. The following are some examples of syntactic parsers.

- The Stanford Parser:

  http://nlp.stanford.edu/software/lex-parser.shtml

- Robust Accurate Statistical Parsing (RASP):

  http://www.informatics.sussex.ac.uk/research/groups/nlp/rasp/project.html

- Collins Parser:

  http://morphix-nlp.berlios.de/manual/node23.html

- Brown parsers:

  http://bllip.cs.brown.edu/resources.shtml#software

Once the question input by the user is parsed into a constituent tree using a syntactic parser, syntactic NLIDB systems use a syntactic grammar to explain users' questions (natural language sentences). An example of the syntactic grammar is as follows (Androutsopoulos et al. 1995).

(1) S → NP VP
(2) NP → Det N
(3) Det → "what" | "which"
(4) N → "rock" | "specimen" | "magnesium" | "radiation" | "light"
(5) VP → V N
(6) V → "contains" | "emits"

The corresponding explanation of this syntactic grammar is as follows:

(1) A sentence consists of a noun phrase and a verb phrase;
(2) The noun phrase consists of a determiner and a noun;
(3) The determiner may be *what* or *which*;
(4) The noun may be *rock*, *specimen*, *magnesium*, *radiation* or *light*;
(5) The verb phrase consists of a verb and a noun;
(6) The verb may be *contains* or *emits*.

Suppose a user's question is

Which rock contains radiation?

According to the syntactic grammar, the NLIDB system could map this question to the following database query (**X** is a variable):

(*for_every* **X** (*is_rock* **X**)
(*contains* **X** *magnesium*)
(*printout* **X**))

One of the best-known syntax-based NLIDB system is LUNAR (Woods 1973). LUNAR was developed to be able to answer questions about moon rock information stored in chemical analysis and literature references databases.

4.3.3.3 Semantic Grammar NLIDB Systems

A semantic grammar NLIDB system is similar to a syntax-based NLIDB system. It also needs a syntactic parser to build the constituent tree of the question sentence input by a user. The constituent tree is then mapped to a database query expression based on pre-defined grammars. The difference lies in the grammar. The mapping from a constituent tree to the database query is based on a semantic grammar in a semantic grammar NLIDB system, rather than a syntactic grammar in a syntax-based NLIDB system (Androutsopoulos et al. 1995). A semantic grammar can be defined to include any semantic concepts according to the user's application. A semantic concept does not necessarily correspond to a syntactic concept (POS tag). An example of semantic grammars is partially shown as follows (Androutsopoulos et al. 1995).

(1) S → Specimen question | Spacecraft question
(2) Specimen question → Specimen Emits info | Specimen Contains info
(3) Specimen → "which rock" | "which specimen"
(4)…

Compared to syntactic grammars, semantic grammars can express richer semantic information which extends the application scope of the system.

We have discussed three basic types of NLIDB systems. The pattern-matching technique is simple and easy to be implemented. Some early NLIDB systems were developed based on this technique and perform well in certain applications. However, the linguistic shallowness of this technique limits its application. The other two types of NLIDB systems have the ability to express more complex patterns for understanding natural language and constructing database queries, but the mapping rules (either syntactic or semantic grammars) are difficult and tedious to devise, which limits the system's flexibility (Androutsopoulos et al. 1995).

In addition to the basic varieties, there are many other types of NLIDB systems. Jung and Lee (2002) developed a multilingual question/answering system using lexico-semantic patterns to support multi-level grammars for query construction. Lexico-semantic patterns are a type of prominent method in text-based question systems. A lexico-semantic pattern contains linguistic information, e.g. words, phrases and POS tags, as well as semantic information, e.g. user-defined classes. Their system was claimed to have high portability of languages, domains, and databases. Domain-specific semantic templates are used to implement a natural language interface to a virtual library by Niculae and co-researchers (2005). A semantic template is a pre-defined pattern and used in runtime to extract the semantic relationships between different objects identified from the input question. Their system includes a pre-processor to build interpretation rules according to the database schema and WordNet [3]. The interpretation rules and semantic templates are then used at run time to extract semantics of input question and construct database queries. More recent development of NLIDB involves a wide variety of other techniques to generate queries for different applications, such as semantically tractable questions (Ana-Maria et al. 2003), domain dictionary and semantic metadata of the database (Rangel et al. 2005), machine learning techniques (Giordani 2008), XML databases (Li et al. 2006), theater database (Roth 2004), video database (MacFadyen 2007) and data warehouses (BusinessObjects 2008; SignaText 2008).

Essentially, the prototype system FACETS developed in this research is under the big umbrella of NLIDB. However, a substantial difference exists between FACETS and traditional NLIDB systems.

NLIDB systems are questions oriented. The user's inputs to a NLIDB system are specific questions corresponding specific answers. For example, an appropriate question accepted by the LUNAR system might be 'How many elements does No. 12 rock include?'. Database queries are constructed based on parsing the user's questions using the NLP technique. Corresponding answers are returned in the form of specific data retrieved from the database based on the constructed queries. The corresponding output of the LUNAR system might be '76 elements'. NLIDB systems, in this sense, are similar to question-answering (Q/A) systems[4].

By contrast, FACETS is decision oriented. Instead of asking questions, managers (users of FACETS) describe and input their current understanding of the decision situation. The user of a NLIDB system usually input one question a time. FACETS is able to accept a complex description of a decision situation consisting of a number of natural language sentences. Some input examples of FACETS are as follows.

"The sales of Mountain-100 Silver, 38 was very impressive. As the successor of Mountain-100 Silver, 38,Mountain-100 Silver, 42 was released in July 2007. Mountain-100 Silver, 42 was designed with higher performance. However, the internet sales of Mountain-200 Silver, 42 are going down in Germany and France since."

---

[3] WordNet is a large lexical database of English, which grouping meaningfully related words and concepts into sets of cognitive synonyms, each expressing a distinct concept. Refer to http://wordnet.princeton.edu/ for more information.

[4] A Q/A system is intended to answer natural language questions by mapping natural language questions to queries into a collection of unrestricted and unstructured texts, instead of queries into a structured database in NLIDB systems.

Furthermore, the processing of each sentence in FACETS is not independent: a sentence is analyzed partially based on sentences that have already analyzed. This mechanism makes it possible to analyze the decision situation as a whole, rather than individual sentences. Instead of definite answers, FACETS presents managers with situation information related to the decision situation. The retrieved situation information is used to support managers to develop situation awareness for decision making. The orientation difference between FACETS and NLIDB demands new models and techniques to analyze the user's natural language inputs.

4.4 Summary

It is a noticeable trend to design DSS with decision makers' cognition at the center of the system, enabling the process of human decision making to be supported from the cognitive perspective. The existing literature shows that managers' SA and mental models have substantial influence on managers' behaviors. Today's business is operated in a complex environment, and is often imposed with a high degree of complexity, uncertainty, dynamics, high personal stakes and time pressure. Successful business decision making heavily relies on managers' SA and mental models. The complex nature of business decision making presents a great requirement of designing DSS in business domain from the cognitive perspective.

Part II
Models

Chapter 5
Cognition-Driven Decision Processes

Human decision making is naturally a cognitive process conducted within a decision maker's mind. However, the use of decision support systems incorporates IS artifacts into the human cognitive process. In this sense, decision making is an integral process comprised of the human cognitive process as well as IS functionality. We use the term *cognition-driven decision process* (CDDP) to denote such a decision making process. In this chapter, a conceptual framework is presented in Section 5.1, which shows high level ideas and essential characteristics of cognition-driven decision processes. Then the detail of the CDDP model is discussed in Section 5.2.

5.1 Essentials of Cognition-Driven Decision Making

5.1.1 The Conceptual Framework of Cognitive Decision Support

The conceptual framework of cognitive decision support (CDS) is the high level 'big picture' of this research as a whole (Figure 5.1). It represents the fundamental idea about cognitive decision support. The CDS framework was adopted from one of our recent publications (Niu et al. 2007).

Essentially, the basic idea expressed in the CDS framework is to facilitate cognitive decision support to business managers (decision makers or general business users) through reinforcement of their situation awareness (SA) and mental models on the basis of information systems. According to the theory of naturalistic decision making (NDM), human decision making is closely related to the decision maker's SA and mental models in complex situations. Richer SA and mental models are more likely to lead to better decisions and eventually result in better decision making performance. The business domain has the potential to apply NDM theory, as business management nowadays becomes increasingly complex and dynamic, with uncertainty, time pressures and high personal stakes.

In Figure 5.1, the dark color box (including *Mental Models* and *Situation Awareness*) in the center denotes a business manager (decision maker). The manager acts as a user as well as an important component of the system. The

L. Niu, J. Lu, and G. Zhang: Cognition-Driven Dec. Supp. for Business Intel., SCI 238, pp. 53–73.
springerlink.com

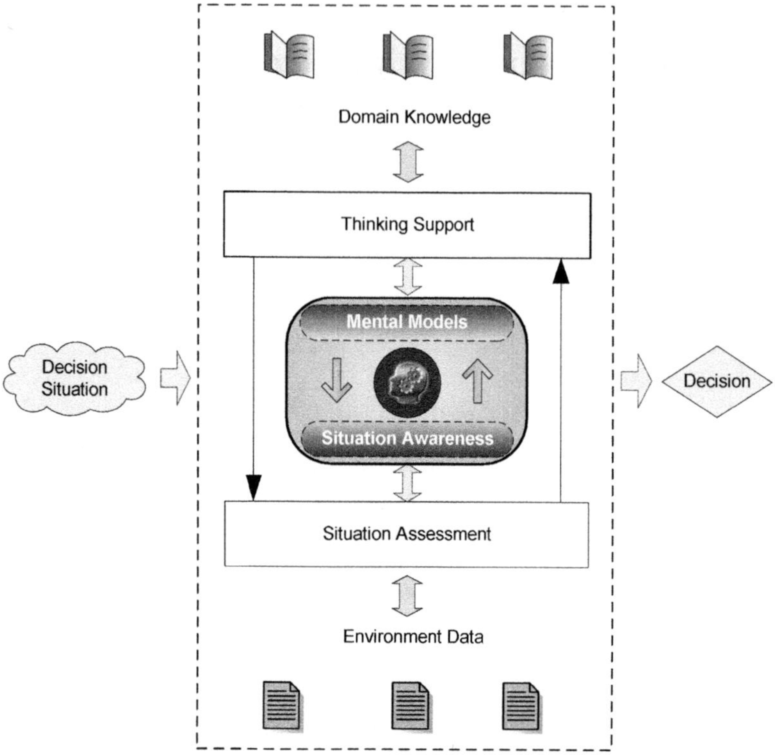

Fig. 5.1 The Conceptual Framework of Cognitive Decision Support

manager's SA and mental models are enriched and rendered by the *Situation Assessment* process and *Thinking Support* process respectively. The interaction between the manager's SA and mental models is a cognitive process; the interaction between *Thinking Support* and *Situation Assessment* is a computerized process. The *Situation Assessment* process and *Thinking Support* process are based on environment data and domain knowledge respectively. The domain knowledge mainly refers to the manager's management experience and the business ontology (Chapter 6). The environment data consists of internal environmental data (e.g., product R&D, financial, engineering, and marketing) and external data (e.g., technological, political, and socio-cultural).

Decision making is problem oriented. A decision problem means something of interest is going wrong or is abnormal and therefore a decision is required to be made and carried out to solve the problem. But in some cases, problems are not obvious or perhaps there are no problems at all. For instance, senior managers are often concerned with identifying potential opportunities or threats before real problems appear. In this book, we use the term *decision situation* to represent a more general meaning than the term *problem*.

Looking at the CDS framework as a whole, we refer to a *situation* (decision situation) as its input and a *decision* as its output. From the IS perspective, the decision situation can be represented as a set of data or information which is referred to as *situation information*. Situation information is either a straightforward description of the decision situation, or indirect information, e.g., the background or other related objects. Generally speaking, a computer-based information system is made up of two basic parts: human and computer. Accordingly, the human-computer system based on the CDS framework receives situation information in two ways: via the human interface (the manager), e.g., participating in meetings, and via the computer interface, e.g., communicating with other systems. The human interface is indicated in Figure 5.1 by the dark color box in the center. The remainder is the computer interface including domain knowledge, thinking support, environment data and situation assessment. We use a dashed box to represent all possible touch points in the interfaces between the human-computer system and its input/output sources.

5.1.2 Cognition-Driven Decision Processes

Compared to traditional BI systems, the decision process in the CDS framework is driven by the decision maker's cognition. Rather than emphasizing behavioral support for the manager's work, the CDS framework focuses on two cognitive aspects of management activities: SA and mental model, and on related cognitive processes, such as thinking, situation assessment and decision formulation. Anthony (1965) categorized management activities into three levels: operational, management and strategic. Correspondingly, we group the manager's SA into three categories: *operational SA, management SA,* and *strategic SA*. At the lowest level, operational SA looks at the most detailed data related to specific tasks usually conducted by frontline personnel as the basis of the operation of the organization. For instance, the awareness of financial transactions and product specifications belongs to operational SA. In the highest level is strategic SA, which reflects aggregated business information such as monthly sale reports and business policies. Management SA is in the middle level. The three levels of SA are relatively roughly defined and there are no absolute boundaries between different levels. For a specific decision making task, the manager might need one or two or all three levels of SA. Take sales as an example. The detailed sales data (reflected by the operational SA) is the major interest of the sales representative. The sales executive is more concerned with aggregated data in sales reports. However the department manager might need both types of data in order to finalize a marketing plan.

The process of SA development and reinforcement is termed *situation assessment* by Endsley (1995b) in terms of cognitive psychology. Situation assessment is a cognitive process which happens in the decision maker's mind. However, in Figure 5.1, we use this term to denote IS functionality (IS based situation assessment) as well as the human cognitive process (cognitive situation assessment). The IS based situation assessment complements and supports the cognitive situation assessment. They both contribute to the manager's SA development and reinforcement for decision making.

The manager's thinking process is supported by the *Thinking Support* component. Thinking is a cognitive process, via which humans are able to recall past experience, manipulate current situation information and then form concepts, reason and make decisions (Jaques 1996). Similar to situation assessment, in Figure 5.1, the manager's thinking process consists of the cognitive process as well as the IS-based process. The former occurs in the managers' mind and the latter is represented by the *Thinking Support* component.

The meaning of thinking support here is twofold. On one hand, the manager's mental models are elicited and then utilized to drive the process of situation information processing. The elicited mental models are stored in a knowledge base. Note the difference between two kinds of mental models: (a) the mental models stored in the knowledge base and (b) the mental models residing in the manager's mind. The former are elicited from the manager, represented and stored as part of the domain knowledge. The latter are concurrent cognitive constructs within the manager's mind regarding past and current decision situations. For the sake of clarity, we refer to the first kind of mental models as experience. Thus, experience in our research is computerized mental models. Both experience and mental models reflect the manager's past management experience and they have important implications for handling the current decision situation. As part of the domain knowledge, experience drives the situation information to be analyzed and processed during a decision process. As cognitive constructs, mental models directly interact with the manager's SA (also cognitive constructs) and drive the decision process to move forward.

On the other hand, the manager's thinking process is supported and enhanced by the information system. People are capable of intuitively using their past experience to handle new decision situations. However, people's ability to reuse experience is subject to many cognitive biases. Some of the cognitive biases can be weakened or eliminated by IS artefacts. In this research, the manager's thinking processes is supported and enhanced through offering the manager interfaces for experience management: input, edit, compare, aggregate, store, retrieve, and graphically present experience.

5.1.3 User Centered Decision Processes

Compared to traditional BI systems, the CDS framework is user centered. User-centered is a design philosophy used for system design and development (Endsley et al. 2003). The opposite is technology-centered design which emphasizes the utilities of technologies in terms of system functionality and the users are required to adapt to the system. User-centered design requires systems to be designed based on not only its fulfillment of functionality but also human factors. The CDS framework is centered on the manager. The major human factors in the CDS framework include the manager's SA, mental models and the related cognitive processes. The manager is concerned with decision making in the decision situation. Thus, the user centered design in the CDS framework can be exemplified by assisting the manager to maker better decisions through facilitating cognitive decision support in specific tasks of the decision process. Some of these tasks in which cognitive decision support is considered in this framework are as follows.

- Describing the Decision Situation

We developed natural language process (NLP) based techniques to offer the manager an interface via which the manager can describe his/her decision situation in the form of natural language (English) (Chapter 7). The NLP based techniques reduce the technical complexity of the current BI interface and aid the manager to more easily develop and enrich his/her SA and mental models for decision making.

- Seeking Relevant Knowledge

Successful decision making requires the manager to obtain sufficient situation information as well as relevant domain knowledge (Section 5.2.1). We developed IS techniques to aid the manager to acquire relevant domain knowledge according to the manager's situation description (Chapters 6 and 7).

- Recalling Past Experience

According to NDM theory, people make decisions based on their past experience. However, the ability of people to recall past experience is subject to many cognitive biases. We developed IS techniques to help the manager to recall past experience and reduce or eliminate cognitive biases (Chapter 6 and 9).

- Obtaining Situation Information

Obtaining the situation information in a timely fashion is the key to situation assessment, during which the manager develops and enriches his/her SA. However, people have an inability to precisely state their information needs. We developed IS techniques to automatically formulate information needs and obtain situation information (Chapter 8).

- Comprehending the Decision Situation

We developed a navigation-knowledge-guided method to present situation information in an intuitive way. This technique is based on the manager's mental models, which enables the manager to perceive and understand the decision situation more comfortably (Chapter 9).

5.2 The Cognition-Driven Decision Process Model

Based on the CDS framework, we model cognition-driven decision processes (CDDP) in Figure 5.2. Part of the work presented in this section has been reported in one of our publications (Niu & Zhang 2008). This CDDP model reflects the conceptual components and procedure of a manager's cognition-driven decision process on the basis of a data warehouse system. Behind this model are three assumptions.

(1) Confronted with a decision situation, the manager will conduct an information processing process, which we called *situation retrieval*. Situation retrieval is motivated by the manager's knowledge need and information need.

(2) The manager is generally unable to state exact information needs for decision-making tasks.

(3) It is possible to derive knowledge need from SA representation.

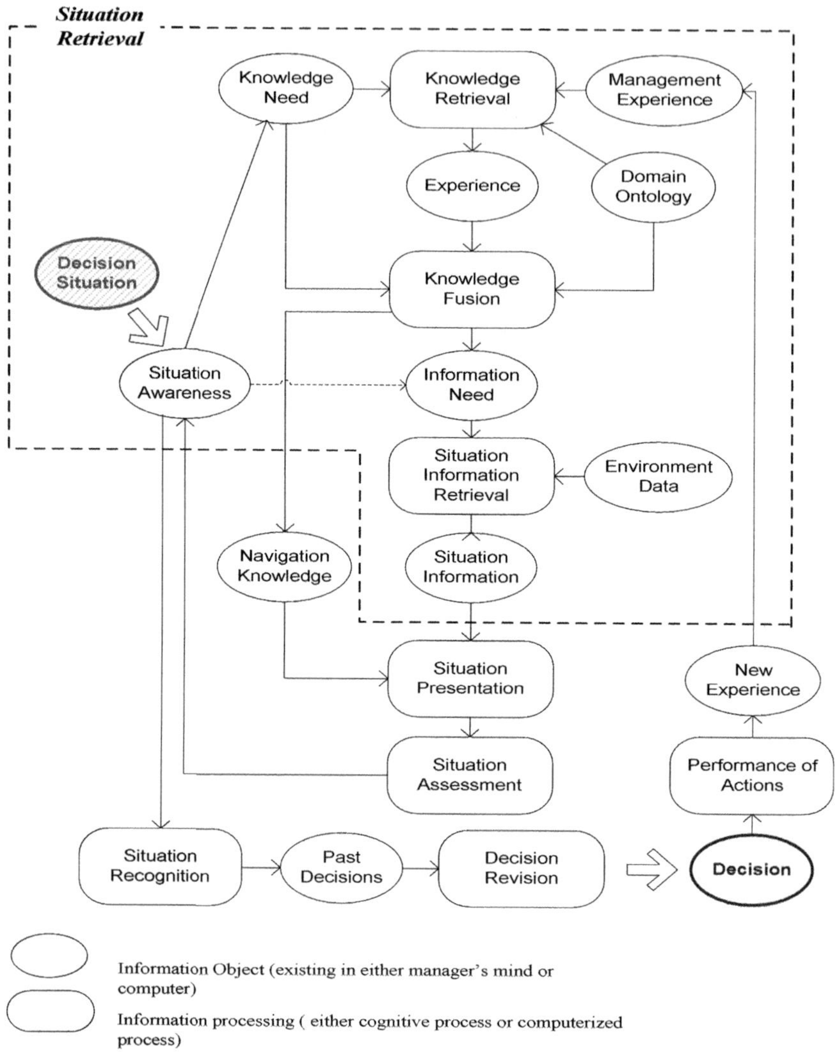

Fig. 5.2 The Cognition-Driven Decision Process Model

The CDDP model is based on the NDM theory, particularly the RPD model. Thus, it requires the manager is an experienced decision maker. The CDDP model describes the cognition-driven decision processes from an IS perspective, which is used to design a new type of BI system with cognitive decision support. Thus, it also requires the acquisition of adequate domain knowledge (stored in a knowledge base including the manager's experience and a domain ontology), and adequate business data (stored in a data warehouse including the environment data).

5.2.1 Situation Retrieval

5.2.1.1 Information Retrieval and Situation Retrieval

We proposed the situation retrieval model based on the information retrieval (IR) model. IR systems are concerned with representing, storing and finding information desired by human users (Ingwersen 1996). Traditional IR systems are based on best-match principle: the matching between documents and statements of the queries put forward by the user (Belkin et al. 1982; Newby 2001). IR systems cater for the users' information needs through searching, locating and obtaining target information. A simplified IR model is shown in Figure 5.3 (Belkin et al. 1982; Ingwersen 1996). With a problem, the user of an IR system will raise a need for relevant information which can be used to solve the problem. The information need motivates the user to interact with the IR system and search for the desired information. Once the user acquires the desired information, the user's problem is solved and the user's information need is dismissed.

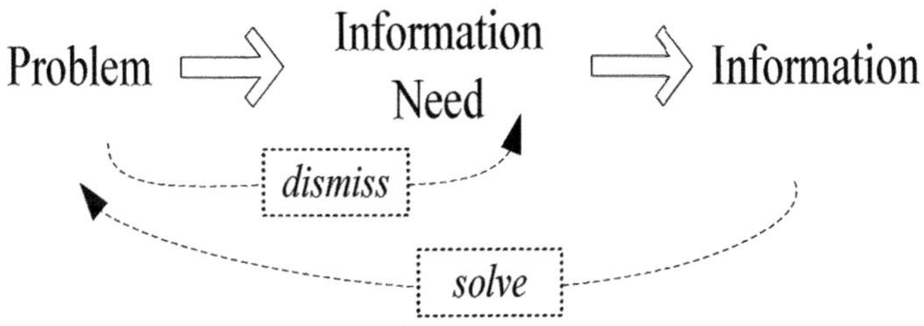

Fig. 5.3 Information Retrieval

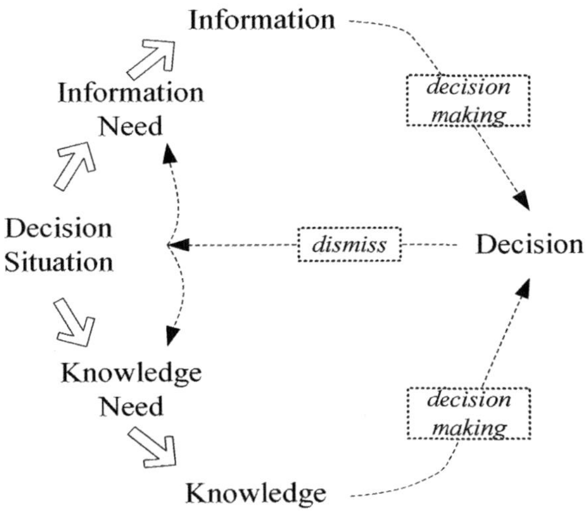

Fig. 5.4 Situation Retrieval

Note: The user is confronted with a decision situation which leads to an information need and a knowledge need. The user consults the system and obtains desired knowledge and information. The obtained information and knowledge is then used to make a decision for the decision situation. A satisfying decision making dismisses the information need and the knowledge need.

We define situation retrieval as the process of searching for situation information as well as situation knowledge for the purpose of decision making. Figure 5.4 illustrates the situation retrieval model. A situation retrieval process begins with a decision situation. The decision situation leads to the decision maker's information need and knowledge need. Respectively, the information need and knowledge need motive the decision maker to seek information (situation information) and knowledge (situation knowledge) relevant to the decision situation for the purpose of decision making. Once a satisfying decision is made and implemented in the decision situation, the corresponding information need and knowledge need are dismissed. The detailed process of situation retrieval is presented in Section 5.2.1.3. There are connections between situation retrieval and IR. According to Figure 5.4, IR is one component of situation retrieval. The objective of IR in situation retrieval is to search for relevant situation information for decision making. Another component is knowledge retrieval which aims at searching for domain knowledge related to the current decision situation. The above discussion leads to the differentiation between data, information and knowledge.

We are not attempting to re-define the concepts of data, knowledge and information here, as many definitions of them have resulted in strong arguments in current literature. However, it would be helpful to differentiate information and knowledge in order to clarify information retrieval and knowledge retrieval, as two basic components of situation retrieval. Based on other researchers' work (Bellinger et al. 2004; Rouse 2002), we roughly distinguish information and knowledge, in that information provides answers to *who*, *what*, *where*, and *when* questions, whereas knowledge implies clues to *how* and *why* questions. In the business domain, information can be specific figures about sales, production, research, marketing trends, and interest rates, which tell the facts about the business. By contrast, knowledge inspires reasoning, explanation and projection based on business facts. The manager's past management experience, business cases, and business procedures fall into the knowledge category.

Information retrieval systems have been very successful in many application areas, such as digital libraries, information filtering and search engines, where users' problems can be solved by offering information as straightforward answers. For example, if a cook wants to know an apple pie recipe, he/she can just simply input keyword 'apple pie recipe' into a search engine and get the desired answer (apple pie recipe details) directly. Therefore, it seems, in IR applications, the target information which users are seeking is all they need for solving their problems (what the cook really needs is the ingredients of apple pies and the cooking procedure). By contrast, knowledge in IR is not as crucial as information for problem-solving (once the cook successfully finds the recipe, his/her problem is solved and extra cognitive processes involving knowledge seem not very necessary).

Unfortunately not all problems are as straightforward as the example of the apple pie recipe. In real life, many problems are very subtle, unstructured and complex, which require extra domain knowledge in addition to sufficient information. In this research, we focus on ill-structured business management problems, e.g., strategic planning. Compared to problems in classic IR applications, business management problems generally require decisions to be made and executed. We refer to this kind of problems as *decision problems*, or more broadly as *decision situations*, as opposed to problems in IR which generally correspond to straightforward answers. In this sense, we say that situation retrieval is decision oriented, while IR is problem oriented. To deal with a decision situation, it is necessary for the decision maker to have relevant information, but information alone is not enough.

Facing a decision situation, the decision maker needs both direct situation information as well as situation knowledge. Situation information is the starting point from which the decision maker develops corresponding SA. In Endsley's SA model (2000), SA is divided into three levels. Seeking and sensing raw situation information is the first step for the decision maker to acquire deep situation understanding. This step results in Level 1 SA. Based on Level 1 SA, the decision maker can develop higher level SA using his/her past knowledge (mental models). However, Level 1 SA cannot guarantee higher level SA as different decision makers have various abilities to digest situation information. This kind of ability depends on decision makers' domain knowledge and on how well they can use their knowledge. Domain experts are differentiated from novices based on their domain knowledge or experience. Mental models are a kind of domain specific knowledge. A mental model is useful in that it provides a mechanism for (Endsley 2000).

(1) guiding attention to relevant aspects of the situation,

(2) integrating information perceived to form an understanding of its meaning, and

(3) projecting future states of the system based on its current state and an understanding of its dynamics.

Therefore, it is possible that a decision maker fails to develop high level SA (Level 2 SA and Level 3 SA) and fails to make good decisions due to his/her limited domain knowledge; even though he/she has already obtained sufficient situation information. For example, many companies possess large amounts of data/information, but do not know (lack of relevant knowledge) how to use them for management. Even after they invest millions of dollars to build data warehouses and enable the data to be easily accessed across the company, the executives still feel lost when facing too many business reports generated by the BI reporting tools (Quinn 2007; Resnick 2003; Sheina 2007b).

The proposed situation retrieval model is to characterize the process of seeking information and knowledge during decision making in a decision situation. In Figure 5.4, the decision maker's knowledge need and information need are provoked by the decision situation. The decision maker raises needs for information and knowledge in order to make an appropriate decision in the current decision situation. Motivated by the knowledge need and information need, the user interacts with the system and gains the desired knowledge and information. Both knowledge need and information need are initiated from the decision situation; however the two constructs also affect each other during the decision process. The more knowledge the user acquires, the more precisely the user is able to know what kind of information he/she actually needs. This results in more accurate and deeper understanding of the decision situation and richer SA of the user, which also positively affect the formulation of the user's knowledge need.

5.2.1.2 Information Need and Knowledge Need

An information need is about what an individual needs to know for problem solving, hence the information need is about what one actually does not know. Confronted with a problem, what one actually knows is that his/her state of knowledge is insufficient (anomalous state of knowledge: ASK) for problem solving which underpins the information need (Belkin et al. 1982). As such, in some cases, although the user can precisely state his/her information need and then proceed to search for corresponding information, in general, the information need is variable, vaguely stated or ill-defined, and difficult to specify (Castells et al. 2007; Salton et al. 1975). Rouse (2002) in particular noticed that users have better abilities to recognize information needs than to specify them. How to support the user to formulate information needs is a classic research question in IR. Nevertheless, a satisfying wide-accepted solution to this question remains under exploration (Cole et al. 2005; Ingwersen 1994; Kuhlthau 1991; Larsen et al. 2006). Due to the unspecifiability of information needs, in Figure 5.2, we use a dotted line to link *Information Need* and *Situation Awareness* to indicate that when the decision maker gains SA, he/she will generate information need implicitly, but cannot state it precisely.

Decision makers' knowledge needs originate from their decision situations, which is a cognitive process in nature. However, in the CDDP model (Figure 5.2), we assume that knowledge needs can be derived from managers' SA representation. Thus in the CDDP model, the decision process will flow from *Situation Awareness* to *Knowledge Need.* The knowledge need directs the manager to the quest for relevant knowledge (knowledge retrieval) and powers the reasoning process for situation comprehension. The role of retrieved knowledge in our model is twofold. On one hand, it helps the formulation of information need; on the other hand, it is used to generate navigation knowledge for situation presentation. SA is the resultant product of situation assessment: the process of seeking, perceiving, and understanding situation information (Endsley 1995b). This process is also affected by SA per se. Sarter and Woods (1991) refer to SA as 'the accessibility of a comprehensive and coherent situation representation which is continuously being updated in accordance with the results of recurrent situation assessments' (p. 52). SA encourages and stimulates the decision maker to actively acquire further situation information and knowledge during decision process (Bergmann 2002; Sarter & Woods 1991). In addition, SA is a state of knowledge (Endsley 1995b). Therefore, it appears that a closed-loop process exists in a situation retrieval process, during which SA, knowledge need, knowledge, information need and information sequentially affect one another (Figure 5.5).

Thus, if we can acquire the statement of a manager's SA, the analysis of the statement will yield representation of his/her knowledge need. In this sense, we set the assumption that knowledge needs can be derived from the SA statement.

Once the knowledge need is derived from the manager's SA statement, the knowledge required for the decision making can be retrieved from the knowledge base. We use the retrieved knowledge to formulate the manager's information need. The formulation of information needs from the knowledge is based on a domain ontology.

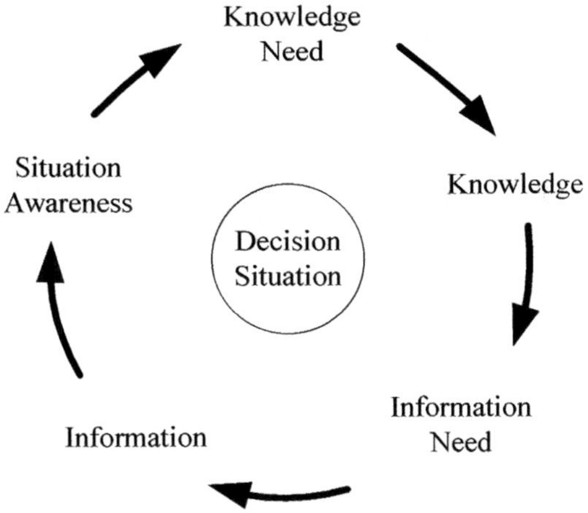

Fig. 5.5 The Loop of Situation Retrieval

5.2.1.3 Situation Retrieval Process

Situation retrieval is a key part of the CDDP model. As Figure 5.2 shows, a situation retrieval process is comprised of five successive steps:

(1) obtaining initial SA (*Decision Situation → Situation Awareness*),
(2) extracting knowledge need (*Situation Awareness → Knowledge Need*),
(3) retrieving situation knowledge (*Knowledge Retrieval*),
(4) generating information need (*Knowledge Fusion*), and
(5) retrieving situation information (*Situation Information Retrieval*).

Step 1. Obtaining Initial Situation Awareness
When confronted with a decision situation, the manager starts the decision process from becoming aware of the decision situation, which results in an initial SA. The manager could obtain an initial SA via conventional communication methods, e.g., business meetings, and the initial SA can be enriched through further communication. The manager's initial SA might be sufficient for some simple decision situations, but complex decision situations might need richer and higher level SA. Hence, in Figure 5.2, following *Situation Awareness* are two alternative routes: *Situation Recognition* and *Knowledge Need*. It is possible that, after gaining the initial SA, the manager is already confident enough to make the final

decision. In this case, the decision process proceeds from the *Situation Awareness* stage directly to the decision generation stage (including *Situation Recognition*, *Past Decisions*, *Decision Revision* and *Decision*). However, in more common cases, the decision process will go through the process of situation retrieval instead of straightforward decision making with the initial SA. The initial SA works as 'seed SA' which triggers the decision process to move forward.

The manager can describe and input his/her SA using natural language. A convenient way to represent SA in the form of natural language is plain text. Thus, a SA can be defined as a set of natural language sentences.

Definition 5.1. Situation Awareness (A)

$$\boldsymbol{A := (S_1, S_2, \ldots S_n)},$$

where, $\boldsymbol{n \geq 0}$, $\boldsymbol{S_i}$ is a natural language sentence.

A SA[5] is an n-tuple consisting of n ordered elements. Each element (S_i) is called a ***SA sentence*** which is a natural language sentence. The following example is a manager's SA consisting of three SA sentences:

Example 5.1. A Manager's Situation Awareness

(*N02 was released in Australia in July 2007.,*

It is a very good product. ,

However, N02 has brought us 60% of notebook sales decline since its release.)

Step 2. Extracting Knowledge Need

By commonsense, when people are communicating with each other, they tend to focus on certain objects in the domain of interest and convey relevant information about these objects in a specific context from a particular view. Consider the following example of SA sentence in Example:

Example 5.2. A SA sentence

"N02 was released in Australia in July 2007."

In this SA sentence, *N02* represents a notebook computer model name. From *N02*, we can infer the following information the manager is trying to deliver.

In the context of **product**, *the manager is talking about* **N02** *which is a specific notebook model name.*

In the context of **product release**, *the manager is talking about* **Australia**, *which is the release location.*

---

[5] According to Collins COBUILD English Dictionary, the word *awareness* is an uncountable noun. However, when a piece of (situation) awareness is represented as an information object, we use this word as a countable noun in this book. This convention also applies to *experience*, *cue*, *ontology* and *navigation knowledge*.

In the context of ***product release****, the manager is talking about* ***July 2007****, which is a release time.*

We define the elementary information units of SA sentences as *SA triples*.

Definition 5.2. SA Triple ($\boldsymbol{T}$)

$$\boldsymbol{T} := (\boldsymbol{c}, \boldsymbol{v}, \boldsymbol{w}),$$

where, $\boldsymbol{c}$, $\boldsymbol{v}$, and $\boldsymbol{w}$ denote the *context*, *view* and *wording* respectively.

A SA triple has three components: *context* (*c*), *view* (*v*), and *wording* (*w*). The wording part represents a word, phrase, or concept, which is the actual information content the SA triple implies. For the SA sentence in Example 5.2, *N02, Australia* and *July 2007* are respectively the wording of three different SA triples.

The wording of a SA triple is understandable only in a specific context. Ambiguity will be introduced without context specification. For instance, the number *2000* does not make sense unless we give it a context, e.g., *sales* (in this context, the sales amount is $2000), or *product release* (in this context, the release year is 2000). We refer to the context of a SA triple as a *native context*. A native context is applicable to the specific wording of a SA triple, i.e., different SA triples might have different native contexts in a SA sentence.

A SA sentence consists of a couple of SA triples. There is also a central context for the entire SA sentence. We refer to the context of a SA sentence as a *local context*. A local context represents the common background where all SA triples in a SA sentence are communicated. For the SA sentence in Example 5.2, *N02* is a product model name. Thus *product* is the native context of *N02*. As for the whole SA sentence, it is talking about a matter of product release. Thus, the local context of this SA sentence can be *product release*. Technically, the local context of a SA sentence should be able to cover all SA triples of this sentence. However, due to the uncertain nature of natural language, it is difficult or impossible to identify such a local context in some cases, e.g., a very vague SA sentence. The method of determining local contexts of SA sentences will be discussed in Chapter 7.

A SA consists of a couple of SA sentences, each of which has a local context. Local contexts might be the same or different from sentence to sentence. Thus, the frequency of each local context can range from *0* to *n* in a SA with *n* sentences. We refer to the overall context of a SA as a *global context*. Let $\boldsymbol{A} = (\boldsymbol{S_1}, \boldsymbol{S_2}, \ldots, \boldsymbol{S_n})$ be a SA. The global context of $\boldsymbol{A}$ is defined as follows.

Definition 5.3. Global Context ($\boldsymbol{c^g}$)

$$\boldsymbol{c^g(A)} := \{\, \boldsymbol{c^l}(\, \boldsymbol{S_i}\,) \mid \boldsymbol{S_i} \in \boldsymbol{A}, \boldsymbol{freq}(\, \boldsymbol{c^l}(\, \boldsymbol{S_i}\,)\,) > \boldsymbol{\kappa} \,\},$$

where $\boldsymbol{S_i}$ is a SA sentence of $\boldsymbol{A}$, $\boldsymbol{c^l}(\boldsymbol{S_i})$ is the local context of $\boldsymbol{S_i}$, $\boldsymbol{freq}(\boldsymbol{c^l}(\boldsymbol{S_i}))$ is the frequency of $\boldsymbol{c^l}(\boldsymbol{S_i})$ and $\boldsymbol{\kappa}$ is the frequency threshold determined by domain experts. The frequency of $\boldsymbol{c^l}(\boldsymbol{S_i})$ is the number of SA sentences whose local context is $c^l(S_i)$. According to this definition, the global context c^g of a SA is a set whose elements are the top frequent local contexts in SA.

In a manager's SA, an object of interest can be described in different ways. For example, *product release* can be described by *release time, release location* and *product model.* The way in which object information is exposed is referred to as ***view*** in Definition 5.2.

Based on the SA triple definition (Definition 5.2), following SA triples can be derived from the SA sentence in Example 5.2:

(*Notebook, Product Model, N02*)

(*Location, Country, Australia*)

(*Time, Time Point, July 2007*)

By the meaning (semantic information) of the SA sentence in Example 5.2, *product release* is an appropriate local context. A native context only applies to a SA triple, while the local context is for the whole SA sentence. Thus, native contexts are narrower contexts compared to local contexts. By Definition 5.1, SA triples are the basic components of SA sentences. To catch the semantic information of a SA sentence as a whole, each SA triple of this SA sentence needs to be re-explained in the local context.

We refer to the process of re-explaining SA triples in a local context as *context propagation.* The context propagation for a SA triple does two things: (a) replaces the native context of the SA triple with the local context; (2) correspondingly revises the view of the SA triple. Applying context propagation to the above three SA triples, three new SA triples are generated as follows:

(*Product Release, Product Model, N02*)

(*Product Release, Release Location, Australia*)

(*Product Release, Release Time, July 2007*)

The context propagation for the first two SA triples is simply replacing their native contexts (*Notebook* and *Location*) with the local context (*Product Release*). For the third SA triple, both its native context and its view are revised.

Context propagation extends the semantic understanding of SA triples from narrower native contexts to wider local context. Thus the sentence context is imposed on each SA triple and richer semantic information of SA can be obtained.

We have discussed the methods to represent SA triples, SA sentence and SA. Now we will look at how knowledge needs are extracted from SA. In the CDDP model, knowledge needs are used to retrieve relevant experience. Experience is elicited from the manager's mental models. Mental models are domain specific and problem oriented (Bergmann 2002; Johnson-Laird et al. 1998). In the RPD model (Klein et al. 1989), mental models act as a kind of pattern about past decision scenarios, which are then compared with the current decision situation by the decision maker.

The manager's SA represents his/her understanding about the current decision situation. In the CDDP model, a SA is comprised of a number of SA sentences, each of which corresponds to a local context. The local context of a SA sentence represents the common background in which the SA sentence conveys actual information. Thus, a local context can be thought of as a sub-topic or sub-problem of the decision situation. There are similarities between local contexts and experience: local contexts represent a pattern of the current decision situation and experience is the pattern of past decision situation. The goal of knowledge retrieval is to get relevant past experience which can be used to aid current decision making. Therefore, it is reasonable to build the knowledge needs based on the local contexts of SA triples in a SA. The following definition shows the method of knowledge need extraction.

Definition 5.4. Knowledge Need (***KN***)

$$\boldsymbol{KN} = \{ \boldsymbol{c}^l(\boldsymbol{S}_i) \mid \boldsymbol{S}_i \in \boldsymbol{A} \},$$

where, $\boldsymbol{A}$ is a SA, $\boldsymbol{S}_i$ is a sentence of $\boldsymbol{A}$, and $\boldsymbol{c}^l(\boldsymbol{S}_i)$ is the local context of $\boldsymbol{S}_i$.

As can be seen from Definition 5.4, the knowledge need corresponding to a SA is a set consisting of the local contexts of all SA sentences. Thus, the knowledge need extracted from a SA is actually equal to the global context of this SA with setting the frequency threshold as zero (0).

Step 3. Retrieving Situation Knowledge
Situation knowledge refers to the manager's management experience which is stored in an experience base. The experience base and the ontology constitute the domain knowledge base. Situation knowledge is retrieved from the experience base according to the knowledge need extracted from the SA.

The retrieved situation knowledge, as the manager's past experience, is reused in two manners. Firstly, the situation knowledge is used for the formulation of the information need. Due to the unspecifiability of information needs, it is difficult to ask the manager directly specify his/her information need. Based on the third assumption of the CDDP model, we use the retrieved situation knowledge to formulate the corresponding information need. Secondly, situation knowledge is used to generate navigation knowledge for situation presentation. A business decision situation usually involves a wide variety and large amount of related data. The way in which the data is presented to the manager heavily affects the process of situation assessment and eventually affects the entire process of decision making. We extract navigation knowledge from the situation knowledge and use the navigation knowledge to guide the process of presenting situation information.

Step 4. Generating Information Need
The manager's information need reflects his/her expectation on the situation information. Situation information is stored in the data warehouse. Thus, information retrieval in our model is equivalent to data retrieval from the data warehouse. Consequently, the information needs can be thought of as queries into the data warehouse. We define information needs as follows:

Definition 5.5. Information Need (*IN*)

$$IN := \{q_1, q_2, \ldots, q_n\}, n \geq 0,$$

where, q_i is a data warehouse query and the execution of q results in an amount of situation information.

According to this definition, the generation of an information need is equal to the construction of relevant data warehouse queries. Therefore, the main problem of information need generation is how to construct appropriate data warehouse queries which can be executed to retrieve the most relevant situation information from the data warehouse for decision making. We developed IS techniques to automatically construct data warehouse queries based on the retrieved situation knowledge and the domain ontology (Chapter 8).

Step 5. Retrieving Situation Information
Situation information is retrieved from the data warehouse according to the information need which is defined as data warehouse queries. Once the information need is generated, queries can be submitted to the data warehouse for execution.

The purpose of situation information retrieval is to find appropriate information relevant to the current decision situation. The retrieved situation information is the basis on which the manager's SA is developed and updated. Thus, it is important to obtain and present the right situation information in a timely fashion. The method of generating information needs ensures the quality of information retrieved from the data warehouse. The retrieved situation information is then presented to the manager based on navigation knowledge.

5.2.2 Generating Navigation Knowledge

The ultimate goal of the CDDP model is to aid managers to make decisions through cognitive support. The manager's SA is developed and enriched in situation assessment. Situation assessment is essentially based on the presentation of situation information. Therefore, the content of situation information as well as the means of information presentation will determine the quality of situation assessment and thus affect the manager's final SA. We have discussed how the situation information is selected and retrieved. Now, we will look at how situation information is presented to the manager.

In current BI systems, there are two basic techniques to retrieve and present information: pre-defined reports and ad hoc querying. Pre-defined reports are based on fixed queries into the data warehouse. The format and content of the report is determined prior to its actual application. The pre-defined reports are 'pushed' to the manager during runtime. Ad hoc querying is mainly used for presenting cube data, where the manager is able to investigate any piece of information on his/her own. As discussed in Section 2.5, the two techniques of information presentation have limitations in that they emphasize manipulation of large volumes of business data, rather than supporting managers' decision making from the cognitive perspective.

Mental models are the mechanism whereby people interact with the outside world. In the business domain, mental models enable managers to simplify the complexity of business environments. Mental models are about people's past experience which are the basis and guidance for adequate SA development. Thus, if situation information can be delivered to the manager in a similar way as the mental models are organized, it might bring the manager some comfort and ease for perceiving and understanding situation information. We refer to the knowledge which can be used to guide situation information presentation as *navigation knowledge*. We developed IS techniques to extract navigation knowledge from the managers' experience.

5.2.3 Situation Presentation

Situation presentation is the visual interface via which the manager interacts with the system and perceives situation information. There are two kinds of information to be presented in this interface: navigation knowledge and situation information.

Navigation knowledge is the manager's integrated experience, which has important implications for handling the current decision situation. People rely heavily on past experience to solve new problems. Therefore, if the situation information can be presented to the manager in the way that navigation knowledge is organized, it will yield a mechanism whereby the manager can perceive information and comprehend the decision situation in an intuitive way. As a result, the manager's cognitive load will be reduced and situation assessment will become an easier and more comfortable mental process.

The navigation knowledge is visualized and presented to the manager in the form of maps consisting of concepts and relationships between concepts. Each concept represents a *cue* implying a possible affecting factor to the current decision situation. A concept is also associated with detailed situation information. Different factors are linked via the relationships. By the links between concepts, different types of situation information are also connected together.

Based on navigation knowledge, a graphical user interface (GUI) is used to communicate situation information to and receive feedback from the user. On the GUI, the navigation knowledge is firstly displayed in the form of maps. The user is navigated within the map via interactive operations. For example, if the manager is interested in a specific concept, he/she can just click on the concept on the GUI and the corresponding situation information associated with this concept will be presented immediately. In this way, the manager can easily browse the situation information of his/her interest and the browsing actions are 'navigated' by the navigation knowledge. The specific technique for navigation-knowledge-guided situation presentation will be discussed in Chapter 9.

5.2.4 Situation Awareness Updating

The manager's SA is updated during the *Situation Assessment* stage in Figure 5.2. Technically, situation assessment is a cognitive process within the manager's mind during which the manager's SA is developed and enriched. However, in the

CDDP model, we use this term to represent IS functionality (IS based situation assessment) as well as the human cognitive process (cognitive situation assessment). The IS based situation assessment complements and supports the cognitive situation assessment. They both contribute to the manager's SA development and reinforcement for decision making.

Broadly speaking, the IS based situation assessment is an ongoing process including all the previous stages of information processing. It starts from the manager's initial SA and goes through knowledge need extraction, situation knowledge retrieval, information need generation, situation information retrieval, navigation knowledge generation and situation presentation. During the IS based situation assessment, different IS techniques are employed to assist the manager to search, locate, obtain and perceive relevant information for the purpose of SA development and enhancement. In this sense, the *Situation Assessment* in Figure 5.2 is not another independent stage of the decision process, but a status of information processing results.

The IS based situation assessment directs situation information to the manager through situation presentation, which triggers the cognitive situation assessment. The cognitive situation assessment, as a cognitive activity in the manager's mind, is divided into three steps (Endsley 1995b). The first step to achieve SA is to perceive the detailed situation information delivered by the *Situation Presentation* stage, e.g., product details, sales amount and stock price. Perceiving this kind of situation information will result in the manager's Level 1 SA.

The second step is to synthesize and comprehend different information perceived in the first step. The goal of decision making and the past experience play an important role in this step. The manager needs to put pieces of information together and form a holistic picture of the current decision situation. The navigation knowledge guided situation presentation can support the second step of the cognitive situation assessment. Some low level situation information has already been aggregated, summarized by the system which helps the manager to easily comprehend the situation information. Different situation information is well organized according to the navigation knowledge, which helps the manager to establish relationships among different elements of the decision situation. The manager's Level 2 SA is developed in this step.

The last step is to project the future status of the decision situation which results in Level 3 SA. The manager's projection ability is based on his/her perception and comprehension of the decision situation (both Level 1 and Level 2 SA). The Level 3 SA enables the manager to predict potential opportunities and threats in the near future, which is valuable for decision making.

5.2.5 Decision Generation

The process of decision generation is based on the recognition-primed decision (RPD) model (Klein et al. 1989) which has been presented in Figure 3.5. Situation assessment results in the manager's updated SA. The updated SA is richer than the initial one developed at the starting point of the decision process. According to RPD model, the user, at this stage, has a better opportunity to recognize the current decision situation and match it with similar past scenarios based on his/her

SA and experience in the *Situation Recognition* stage in Figure 5.2. The past decisions are then mentally examined. Depending on the appropriateness, past decisions might be adopted directly for the current decision situation, or be revised accordingly. A feasible decision to the current decision situation is made through revising the past decisions.

Compared to the traditional DSS-based decision process model, the final decision, in the CDDP model, is made by the manager him/her self, instead of a decision recommendation generated by the computer. In other words, only *decision support* is provided to the manager in the form of reinforcement of SA and mental models, and enhancement of situation assessment and thinking.

Decision making is also a learning process, during which the decision maker gradually accumulates field expertise. Once a course of action is generated and implemented after decision making. The decision maker has opportunities to gain new experience from the performance of a course of action. The experience base will grow more reliable through constant intake of new experience.

5.2.6 The Decision Cycle

Subject to many factors, the manager may not make the final decision after going through the *Situation Recognition* process. For example, the manager feels his/her SA is not sufficient enough for decision making. In this case, the manager will go through another iteration of the situation assessment through re-inputting the updated SA. Therefore, a cognition-driven decision process might include a number of iterations of decision cycle. As shown in Figure 5.6, each decision cycle consists of a series of six consecutive steps corresponding to the different stages in the CDDP model.

(1) SA Analysis

The manager describes his/her SA in natural language. The SA is analyzed in order to extract knowledge needs. The SA being analyzed is either the manager's initial SA or updated SA. The initial SA can be developed via conventional communication methods and the updated SA is the result of situation assessment.

(2) Knowledge Retrieval

Experience is retrieved from the experience base according to the knowledge needs extracted from the manager's SA in Step (1). The retrieved past experience is relevant to the current decision situation.

(3) Knowledge Fusion

The retrieved knowledge is synthesized to form information needs and navigation knowledge.

(4) Situation Information Retrieval

Situation information retrieval is conducted based on information needs generated in Step (3). The goal of situation retrieval is to retrieve information which is relevant to the current decision situation.

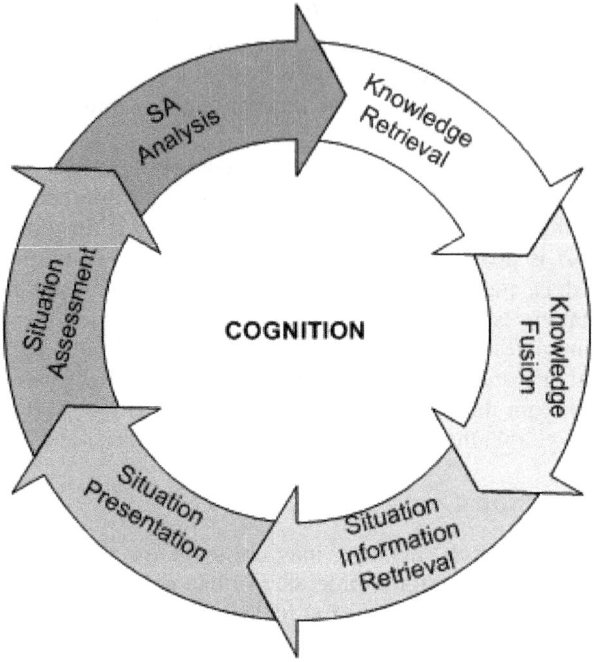

Fig. 5.6 The Decision Cycle of Cognition-Driven Decision Processes

(5) Situation Presentation

The retrieved situation information is presented to the manager under the guidance of navigation knowledge. Situation information presentation is the basis of situation assessment.

(6) Situation Assessment

The manager perceives situation information through interaction with the system and his/her SA is eventually developed and enriched.

The iteration of decision cycle is triggered by the manager's initial SA. After that, the decision cycles iterate one after another until the final decision is worked out. There are two cases in which the final decision is made. First, during an iteration of decision cycle, the manager feels confident enough at a point in time to finalize a decision. The manager's confidence mainly relies on his/her mental models and SA. Solid mental models and high level SA are more likely to lead to successfully decision making. Second, limited resources available to the manager might force the manager to make a decision. Examples of these resources include time, cost, and the manager's cognitive load. Any resource limits may stop the iteration of decision cycle and trigger the production of the final decision.

5.3 Summary

The CDS framework and the CDDP model are presented in this chapter. The CDS framework outlines two basic characteristics of a cognition-driven decision process: cognition oriented and user centered. According the CDS framework, the CDDP model was proposed based on three assumptions. The key part of the CDDP model is situation retrieval. The major difference between situation retrieval and classic information retrieval is that the former is decision oriented and the latter is problem oriented. The goal of a situation retrieval process is to seek, locate and obtain knowledge and information relevant to a decision situation. Situation retrieval is the fundamental part of a cognition-driven decision process comprised of iterations of decision cycle. During a decision cycle, the manager's mental models and SA are represented as computerized information objects and utilized for formulation of knowledge needs and information needs, knowledge retrieval, situation information retrieval and situation presentation. In this sense, the manager's decision process is driven by his/her cognition and cognitive decision support is achieved. Although the CDDP model technically is domain-independent, the discussion of this model in this book is grounded on the business domain for the sake of clarity. We believe the CDDP model can be easily extended into other domains, such as government, medical and education.

This chapter represents the theoretical part of this cognition decision support for BI. The corresponding technical part, including methods, algorithms, and techniques developed on the basis of the CDDP model, will be discussed in Chapters 6, 7, 8 and 9.

Part III
Techniques

Chapter 6
Domain Knowledge Representation and Processing

From this chapter, we present the technical part of this book including Chapters 6, 7, 8 and 9. The technical part deals with problems related to specific methods, algorithms, and techniques which are the implementation of the cognition-driven decision process (CDDP) model. This chapter discusses two kinds of domain knowledge used in the CDDP model: ontology and experience. A domain ontology is represented as a class tree and a class graph based on two types of relationships between classes: subsumption and property-share. Experience is elicited and represented as experience maps using the cognitive mapping technique. The domain ontology is used for extracting semantic information of SA. The extracted SA semantics together with experience are then used for query construction. This chapter will focus on domain ontology and experience.

6.1 Ontology

6.1.1 Basics of Ontology

Ontologies are originally used to offer a basis for communication between different parties in order to dismiss ambiguities (McGuinness 2001). However, recent research has brought ontologies to much broader areas: battlefield (Matheus et al. 2003) semantic web (Rocha et al. 2004), information extraction (Boufaden 2003; Maynard et al. 2005), business management (Missikoff & Schiappelli 2005), and information retrieval (Castells et al. 2007). In particular, Uschold and colleagues (1998) developed an ontology (*The Enterprise Ontology*) for business domain. They summarized the application of the ontology into four major roles:

(a) a communication medium,
(b) acquisition, representation, and manipulation of enterprise knowledge,
(c) structuring and organizing libraries of knowledge, and
(d) explanation of the rationale, inputs and outputs of the software tools.

In computing science, an ontology is a specification of data model about classes (concepts) and their relationships (Guarino 1998). In an ontology, a class is a word or phrase that expresses a general idea of the nature of a thing, often providing a

L. Niu, J. Lu, and G. Zhang: Cognition-Driven Dec. Supp. for Business Intel., SCI 238, pp. 77–96.
springerlink.com

category for the classification of things (Theodorson & Theodorson. 1969). Classes are domain-specific. In the business management domain, examples of classes are STRATEGY, MARKETING, PRODUCT, SALES and CUSTOMER[6]. A class is differentiated from other classes according to their properties, i.e., different classes have different sets of properties. For instance, the property set of CUSTOMER might include Name, Age, Education, and Income.

In an ontology, different relationships exist between classes. Subsumption is a very basic type of class relationship. An example of subsumption relationship is the one between PRODUCT and NOTEBOOK. NOTEBOOK belongs to PRODUCT: any individual (instance) of NOTEBOOK is also a member of PRODUCT. In a specific domain, additional types of relationship can be defined in order to meet different semantic requirements. For instance, there is *wrote* relationship between AUTHOR and BOOK, and *is-topped-with* relationship between PIZZA and TOPPING. It is noticeable that a hierarchical taxonomy can be defined if only subsumption relationships between ontology classes are depicted. In addition to the subsumption relationships, we defined a new type of relationship called ***property-share relationships*** for knowledge needs construction.

6.1.2 Property-Share Relationships

In real life, people learn of things via obtaining information about them. The information can be either direct or indirect, subject to the observation method. Observing a thing directly, people can obtain direct information. Indirect information can be obtained via intermediaries. For example, people can image how a child (the target) looks like according to his/her parents' (the intermediary) appearance. Indirect information can be considered as a sort of relationship between the intermediary and the target. The information conveyed by this relationship is actually based on the properties of the intermediary which are shared with the target. This information is the intermediary's contribution to the understanding of the target. In other words, we can know something about the target via examining the shared properties of its intermediaries. For the previous example, the child shares some properties (e.g., facial appearance or personal nature) with the father. We refer to this kind of relationships between ontology classes as ***property-share relationships***.

A property-share relationship between two classes is a relationship directed from the intermediary class to the target class. Let $\boldsymbol{A}$ and $\boldsymbol{B}$ be an intermediary class and a target class respectively. Let $(\boldsymbol{a_1}, \boldsymbol{a_2}, \ldots \boldsymbol{a_m})$ be properties of class A and $(\boldsymbol{b_1}, \boldsymbol{b_2}, \ldots \boldsymbol{b_n})$ be the properties of class $\boldsymbol{B}$. The property-share relationship from $\boldsymbol{A}$ to $\boldsymbol{B}$ is defined as follows:

Definition 6.1. Property-Share Relationship ($\boldsymbol{r^s}$)

$$\boldsymbol{r^s}\,(\boldsymbol{A}, \boldsymbol{B}) := \{ <\boldsymbol{a_i}, \boldsymbol{b_j}, \boldsymbol{f_i}> \mid \boldsymbol{a_i} \in \boldsymbol{A}, \boldsymbol{b_j} \in \boldsymbol{B} \textbf{ and } \boldsymbol{f_i}(\boldsymbol{a_i}) = \boldsymbol{b_j} \}.$$

---

[6] In this book, ontology classes (concepts) are presented in full upper-case (e.g., MARKETING), while the properties of a class are only capitalized (e.g., Sales Amount).

In this definition, f_i is a ***property transformation function*** which maps a property ($\boldsymbol{a_i}$) of $\boldsymbol{A}$ to a property (Carley & Palmquist 1992; Gnyawali & Tyler 2005) of $\boldsymbol{B}$. f_i is defined according to specific application requirements. In most cases, $f(p) = p$ can be adopted, i.e., the target class and the intermediary class share properties directly without any transformation. Let us look at an example to illustrate different property transformation functions.

PRODUCT RELEASE = (Release Date, Product ID)

SALES = (Start Date, End Date, Product Model)

r^s (PRODUCT RELEASE, SALES) = {<Release Date, Start Date, f_1>, < Product ID, Product Model, f_2>}, where $f_1(p) = p + N$, N is a period of time, and $f_2(p) = p$.

With the property-share relationship r^s, the Start Date of SALES can be obtained from Release Date of PRODUCT RELEASE by applying property transformation function f_1.

According to property transformation functions

$f_1(p) = p + N$,

Start Date = Release Date + **N**.

Suppose

N = 5 days.

Thus, given a date when a product is released, a reasonable start date of this product's sales can be determined. For instance, if we know the product was released on 12 June 2008, then it is reasonable to examine the sales data of this product from 15 June 2008. The Product Model of SALES can be obtained from Product ID of PRODUCT RELEASE by property transformation function f_2. As $f_2(p) = p$, Product Model = Product ID. Thus, r^s enables us to gain insights into the sales of a product through available data about its release, even we do not have direct sales data at hand. Well-defined property-share relationships have broad implications in business management, as they provide us with a means to perceive useful information indirectly.

With reference to the definitions of ontology relationships, we defined two data structures to represent a domain ontology: ***class tree*** and ***class graph***. Accordingly, the class tree and class graph of an ontology respectively present the subsumption relationships and property-share relationships between ontology classes. Note that the class tree and class graph are NOT two different ontologies, but two representations of the same ontology from two different perspectives.

As the proposed theory, methods and algorithms in this research are not limited to a specific domain, the presentation of ontology is also not domain specific. However, most examples (classes and relationships) used in this section are excerpts from the ontology developed in the application case studies of this research (Chapter 11).

6.1.3 Class Tree

The building blocks of class trees are nodes and links. A node of a class tree denotes a class in the ontology. The subsumption relationship between two ontology classes is represented as a link in the class tree. Figure 6.1 is an excerpt of the of class tree developed in the business application case study of this research (Section 11.1). The structure of a node in a class tree has seven fields.

(1) Class Label

The class label of a node is the name of the class that node represents. A class label is an identification of a class. Therefore, class labels are unique in a specific ontology (including its class tree and class graph). However, the reality is that there are many more classes used in an application domain than in the ontology defined for that domain. In addition, different classes might mean a similar thing, or the same class might mean different things in different contexts. In order to avoid ambiguity of communication, all classes in an ontology must meet the uniqueness requirement. Thus, the determination of all classes for an ontology is by no means an straightforward task. Ontology development often involves extensive cooperation of domain experts, technical experts and engineers.

(2) Parent Class

A class tree only represents the subsumption relationships between classes. Therefore every class (node), except for the root, has a corresponding parent class (node). The parent class field of a class node is also a class label.

(3) Class Properties

A class of an ontology is defined by a set of properties. The distinction between the property set of a class and a node structure should be noted. A node of a class tree is an artifact designed to represent a class in the form of a data structure (tree node). The properties of a class are the natural description of the class (existence).

Of the property set of a class, we define a special one as ***taxonomy property***. The taxonomy property of a class is used to distinguish its child classes. For example, Product Category is the taxonomy property of PRODUCT. PRODUCT has four children: BIKE, COMPONENT, CLOTHING and ACCESSORY. These four child classes are differentiated from each other by the Product Category property. Although a child class inherits all properties from its parent class, still the child class might have a different taxonomy property to its parent class.

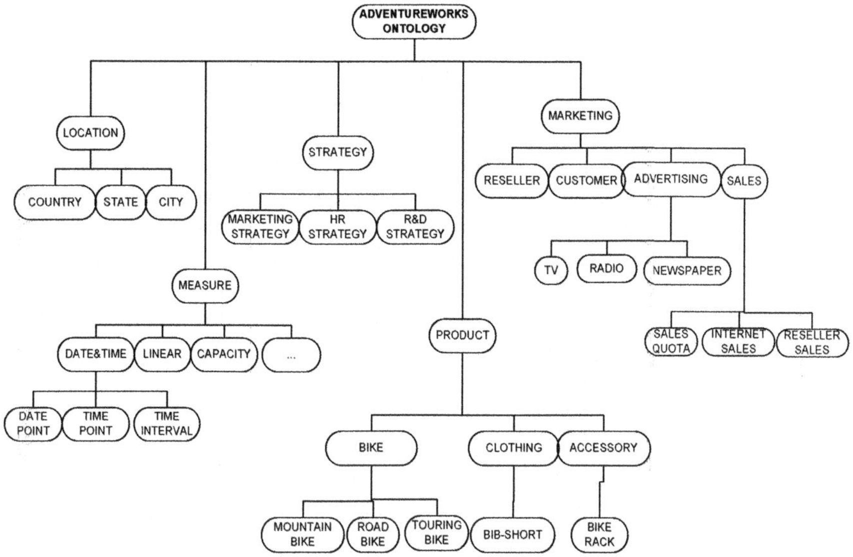

Fig. 6.1 An Example of Class Tree

(4) Class Type

A class can be one of three class types: ***general class***, ***abstract class*** and ***meta class***. In the hierarchy of a class tree, general classes are the lowest level classes (leaf nodes). General classes can have specific instances. An instance of a general class can be uniquely identified by the taxonomy property of the class. Thus, a value of the taxonomy property of a general class implies the presence of an instance of this class and then implies the presence of this class itself. For example, English Product Name is the taxonomy property of MOUNTAIN BIKE. Any valid value of English Product Name, e.g., *Mountain-100 Silver* represents a specific mountain bike (an instance). This instance will imply the presence of class MOUNTAIN BIKE.

Compared to general classes, abstract classes are higher level classes in a class tree. Abstract classes do not have direct instances. Therefore, abstract classes cannot be identified by instances. We use semantic parsing to deduce abstract classes from managers' SA descriptions (Chapter 7).

Meta classes are auxiliary classes which are used to characterize general classes and abstract classes. A meta class has only one property: value. A meta classes is usually a property of a general or abstract class. For example, MONEY may correspond to Sales Amount which is a property of SALES.

There are two types of meta classes: ***basic meta classes*** and ***extended meta classes***. The basic meta classes are pre-defined and not connected to specific application domains. Currently, we have defined 16 meta classes: AREA, SPEED, CURRENCY, TEMPERATURE, DISTANCE, TIME INTERVAL, TIME POINT, LENGTH, HEIGHT, VOLUME, POWER, WEIGHT, PRESSURE,

COUNTRY, STATE, and CITY. Clearly, the pre-defined meta classes can only describe a limited number of types of classes for a specific domain. Thus, it is reasonable to allow users to define their own meta classes according to their requirements. The user-defined meta classes are domain-specific and we refer to them as extended meta classes.

The number of general and abstract classes in a specific domain can range from tens to over one thousand. Each class has a couple of properties. It might be labor-intensive and time-consuming to manually define all classes for an application. Fortunately, we created a mapping relationship between the ontology and the data warehouse. Thus each class in the class tree is associated with a table in the data warehouse. The classes can be easily defined by importing data warehouse schema.

Both general classes and meta classes can have specific instances. For the sake of distinction, we refer to instances of meta classes as ***meta instances***.

(5) Is Experience Map Concept?

In order to apply the CDDP model in an application, an ontology and a corresponding experience base needs to be built. An experience base includes the managers' management experience represented as experience maps. Each experience map consists of concepts and causal relationships.

The fifth field of a class node indicates whether this class is also a concept in the experience base. We map every concept of the experience maps to a class in the ontology. However, not all ontology classes correspond to a concept in the experience maps due to the scope difference between the ontology and the experience maps. Theoretically, an ontology is intended to represent all classes of interest and their relationships in an application domain, i.e., the problem space. In comparison, an experience map as the representation of a manager's experience is a subset of the solution space. In terms of applicability, the ontology has a wider scope than the experience map. Consequently, the concept set of experience maps is a subset of the class set of the ontology.

(6) Data Warehouse Object Type

This field together with the seventh field specifies an object in the data warehouse corresponding to this class. By this means, the mapping relationship between an ontology and a data warehouse is established. This mapping relationship is used to construct data warehouse queries (Chapter 8).

We are concerned with three kinds of data warehouse objects: data table, dimension tables and cubes. Data tables contain relational data the same as tables in a relational database management systems (RDBMS). Cubes store multidimensional data which are defined by a number of dimensions. Each dimension is built based on a number of dimension tables. We use *0*, *1*, and *2* to denote data tables, dimension table and cubes respectively. More detailed discussion about data warehouse tables and cube is presented in Section 8.1.1.

(7) Data Warehouse Object

This field is the name of the data warehouse object mapped with this class. For example, in the ontology for the business application case study (Chapter 11), the corresponding data warehouse objects of PRODUCT and INTERNET SALES is *DimProduct* table and *Internet Sales* cube respectively. Technically, all classes should have corresponding objects in the data warehouse. However, due to the limited scope of a specific application, it is not necessary to relate every class to a data warehouse object.

6.1.4 Class Graph

A class graph is another part of the representation of an ontology in addition to its class tree. Compared to the subsumption relationships in the class tree, property-share relationships are represented in the class graph. The class graph is a directed graph representing the same set of classes as the class tree. Each vertex denotes a class; each directed edge denotes a property-share relationship (Figure 6.2).

The class tree of an ontology includes all classes defined for an application domain, but not every class defined the class tree has property-share relationships with other classes. Only those class pairs involved with property-share relationships are represented in the class graph. In practical applications, users only need to define necessary property-share relationships of their interest, instead of all possible class pairs.

The example in Figure 6.2 is an excerpt of the class graph for the business application case study of FACET (Chapter 11). There are two types of edges in this class graph: monodirectional lines and bidirectional lines. A monodirectional edge denotes a property-share relationship, e.g., the edge from PRODUCT SUBCATEGORY to PRODUCT denoting r^s(PRODUCT SUBCATEGORY, PRODUCT). A bidirectional edge denotes two property-share relationships, e.g., the edge between SALES and CUSTOMER denoting r^s(SALES, CUSTOMER) and r^s(CUSTOMER, SALES). As shown in Figure 6.2, most property-share

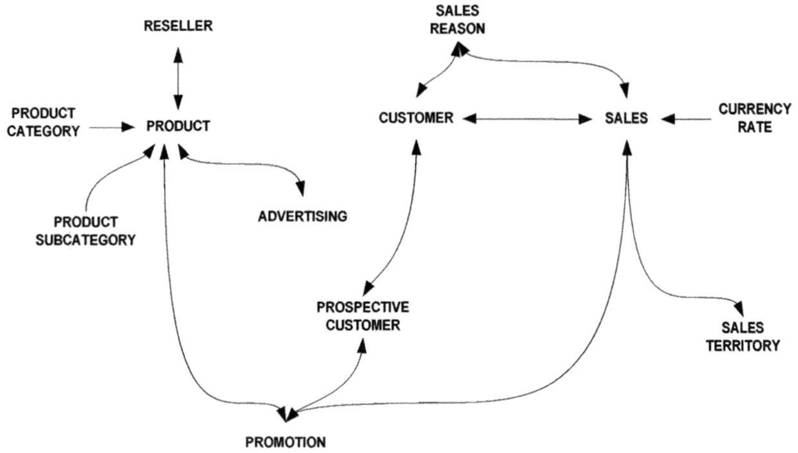

Fig. 6.2 An Example of Class Graph

relationships are symmetric, i.e., both r^s(A, B) and r^s(B, A) are valid property-share relationships, e.g., r^s(PRODUCT, ADVERTISING) and r^s(ADVERTISING, PRODUCT).

6.1.5 Role of the Ontology

The major role of the domain ontology (its class tree and class graph) in the CDDP model is to provide a mechanism whereby the semantic information of SA can be extracted.

(1) Annotates SA terms

In the CDDP model, managers can describe their SA using natural language. The SA is represented as texts, each of which includes several sentences. Employing a natural language parser, SA sentences are parsed into syntactic tokens (SA terms). SA terms are words or phrases constituting a SA sentence. SA terms are annotated based on an ontology. The annotated SA terms form SA triples (Chapter 7).

(2) Infers abstract classes

In terms of subsumption relationships, abstract classes are at the higher level in a class tree compared to general classes (leaf nodes). General classes can be directly extracted from SA descriptions according to their taxonomy properties. Abstract classes generally need to be inferred from abstract classes and relevant SA terms based on the subsumption relationships (Chapter 7).

(3) Determines class similarities

Class similarities represent the degree to which two classes are talking about the same thing. Class similarities are determined by comparing classes in both the class tree and the class graph (Chapter 6).

(4) Represents the relationships between objects in the data warehouse.

Situation information is stored in the data warehouse, which can be retrieved according to information needs. The construction of information needs relies on a deep understanding of the relationships between different data objects in the data warehouse, such as relational tables, dimensions and cubes. We map each data warehouse object of interest to a class in the corresponding ontology. Thus, relationships between classes in the ontology can reflect the relationships between objects in the data warehouse.

6.1.6 Synonyms

The synonyms of a class are terms which have the same or very similar meanings as the class. Using synonyms, the communication among people or information systems can be based on wider vocabulary than those used in an ontology. Thus, synonyms extend the applicability scope of an ontology and give practical applications more freedom to communicate using conventional terms.

Synonyms can be generated automatically based on thesauruses defined in a specific domain. Users can also manually define as many synonyms as they want. In FACETS, synonyms are a list of words in a text file. Each entry in the file consists of all synonyms defined for the same class. For instance, the entry for STRATEGY is defined as follows:

strategy, strategies, plan, plans, planning, planned, policy, policies, procedure, procedures

The synonyms of a class include nouns, verbs and different forms of these words.

6.1.7 Class Similarity

Class similarities are used to retrieve an alternative class when the required class cannot meet the specified criteria. For example, in constructing data warehouse queries, a target data warehouse object corresponding to a class needs to be determined for a given concept in a cue. If the class matched with the concept does not have a corresponding data warehouse object, the most similar class with a data warehouse object needs to be determined.

In an ontology, the similarity between two classes is computed based on the class tree and the class graph of this ontology. First, their similarity is evaluated in the class tree (***tree similarities***) and class graph (***graph similarities***) individually and then aggregated together to form the class similarity.

- Tree Similarities

The class tree of an ontology is a hierarchical structure. The methods to computing class similarities within a hierarchical ontology can be grouped into two basic categories: the edge based methods and the node based methods (Schickel-Zuber & Faltings 2007). Schickel-Zuber & Faltings compared various existing methods and concluded that there are only marginal differences between different methods in terms of mean absolute error measure. This conclusion encourages us to adopt a relatively simple similarity evaluation method as shown in following equation. This method was proposed by Leacock & Chodorow (1998).

Equation 6.1. Tree Similarity (***ts***)

$$ts(a,b) = -\lg\left(\frac{len(a,b)}{2D}\right),$$

where, ***a*** and ***b*** are two nodes of a class tree and ***len***(***a***, ***b***) is the shortest length of the path from node ***a*** to node ***b*** in the class tree. The length of a path in the class tree is defined as the number of nodes in the path. *D* is the maximum depth of the class tree.

- Graph Similarities

A class graph represents the property-share relationships between ontology classes. Let ***a***, ***b*** be two classes in a class graph and $r^s(a, b)$ be the property-share relationship from ***a*** to ***b***. The properties of ***a*** defined in $r^s(a, b)$ can be transformed into corresponding properties of ***b*** by a transformation function. The more properties of ***a*** are transformed to ***b***, the stronger $r^s(a, b)$ is. Therefore, the property-share relationships, to some extent, reflect the degree to which classes are related to each other: the stronger a property-share relationship between two classes, the closer the two classes. Thus, we defined graph similarities using equation

Equation 6.2. Graph Similarity (***gs***)

$$gs(a,b) = \prod_{edge(c_i,c_j) \in path(a,b)} \left(\frac{|r^s(c_i,c_j)| + |r^s(c_j,c_i)|}{2|c_i|} \right).$$

In this equation, ***gs***(***a***, ***b***) is the graph similarity from **a** to **b**. ***path***(***a***, ***b***) is the shortest path from vertex ***a*** to ***b***. ***edge***(c_i, c_j) is one of the series of edges in ***path***(***a***, ***b***). $r^s(c_i, c_j)$ and $r^s(c_j, c_i)$ is the property-share relationship from c_i, to c_j and from c_j to c_i, respectively.

To combine the tree similarity and the graph similarity, the class similarity between two classes can be computed using the following equation.

Equation 6.3. Class Similarity (*sim*)

$$sim(a,b) = \mu \cdot ts(a,b) + (1-\mu) \cdot ts(a,a^s) \cdot ts(b,b^s) \cdot gs(a^s,b^s).$$

In this equation, μ ($0 \leq \mu \leq 1$) is the weight of the tree similarity which is determined by experts according to specific applications. a^s, b^s are respectively the ***surrogate classes*** of ***a*** and ***b***. In a class graph, the surrogate class of a class ***c*** is the one that has the greatest tree similarity with ***c*** compared to all other classes of the class graph. As discussed in Section 0, the class graph of an ontology only includes part of classes defined in the class tree. Those classes not involving in property-share relationships will not appear in the class graph, e.g., most meta classes. For each of these classes, a surrogate class is used to compute the graph similarities with other classes.

6.2 Experience

A decision maker, e.g. a business manager, acquires and accumulates relevant domain knowledge during decision making processes. As a kind of such domain knowledge, mental models play a key role for SA development (Endsley 1995b; Sarter & Woods 1991). Mental models can be elicited using the cognitive mapping technique. We refer to computerized mental models as *experience*.

6.2.1 Experience Representation

Based on related work on cognitive mapping, we defined experience and related basic concepts for experience representation.

Let $\boldsymbol{C} = \{c_1, c_2, \ldots, c_n\}$ be a concept set. Concepts c_i, c_j belong to $\boldsymbol{C}$. The causal relationship from c_i to c_j is defined as follows.

Definition 6.2. Causal Relationship (r^c)

$$r^c\,(c_i, c_j) := (c_i, c_j), \text{ where } c_i \in C,\ c_j \in C,\ c_i \neq c_j, \text{ and } c_i \text{ is a cause of } c_j.$$

A causal relationship defines an ordered pair of concepts, of which the former concept (c_i) is a cause of the latter concept (c_j). We refer to c_i as a ***cause concept*** of c_j and accordingly c_j as an ***effect concept*** of c_i.

In a causal relationship, the cause concept is a ***direct*** cause of the effect concept. The cause concept directly affects the effect concept in some ways without any intermediaries. ***Indirect*** causes do not guarantee causal relationships between concepts. For instance, if we have $r^c\,(a, b)$ and $r^c\,(b, c)$, we can conclude that a is a cause of c. However $r^c\,(a, c)$ does not necessarily exist. In this case, a is an indirect cause of c. We refer to indirect causal relationships as *mediated causal relationships.*

Let C be a concept set. We give the definition of mediated causal relationships as follows.

Definition 6.3. Mediated Causal Relationship (r^m)

$$r^m(a, b) := (a, b),$$

where, $a \in C,\ b \in C,\ a \neq b$,

and, $\exists C' \subseteq C,\ C' = (x_0, x_1, x_2, \ldots, x_p, x_{p+1}),\ p \geq 1,\ x_0 = a,\ x_{p+1} = b,$

such that $\forall x_i \in C',\ 0 \leq i \leq p,\ r^c(x_i, x_{i+1})$ is a causal relationship.

According to this definition, a mediated causal relationship from one concept to another concept implicitly defines an ordered tuple of concepts between the two concepts. A valid causal relationship exists for each pair of adjacent concepts.

Based on Definition 6.2, an experience can be defined as follows.

Definition 6.4. Experience E

$$E := (\,C,\ R\,),$$

$$C = \{\ c_1, c_2, \ldots, c_n\ \},\ c_i \text{ is a concept of concern},$$

$$R = \{\ r^c(c_i, c_j) \mid c_i \neq c_j,\ c_i \in C, c_j \in C\ \},$$

where, $r^c(c_i, c_j)$ is a causal relationship directed from c_i to c_j.

According to Definition 6.4, an experience can be represented by two parts: a set of concepts and a set of causal relationships between concepts. A concept represents an entity or object of an individual's concern. Examples of concept in

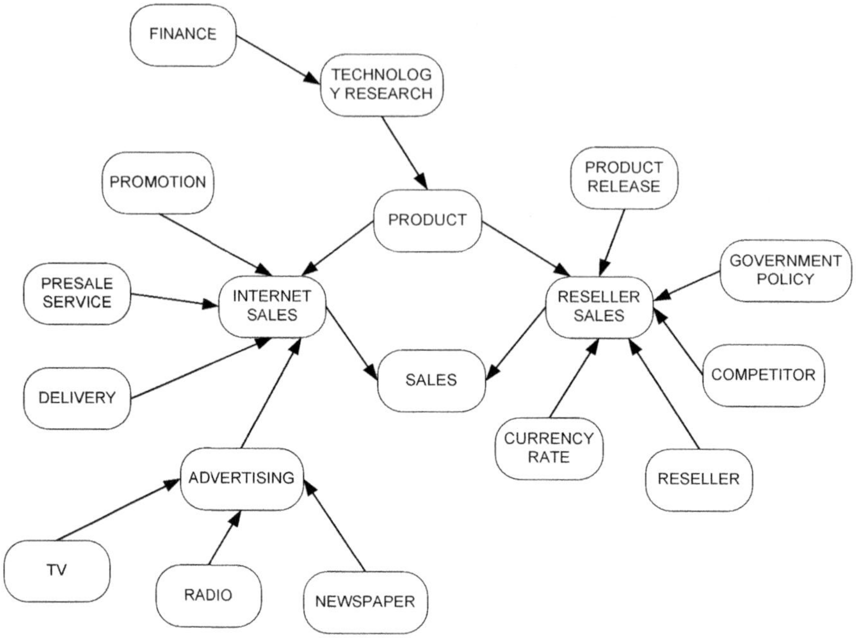

Fig. 6.3 An Example of Experience Map

the business domain are SALES, PRODUCT RESEARCH, MARKETING, and SHARE PRICE. The causal relationship between two concepts in an experience is required to be direct. For example, PRODUCT RESEARCH directly affects SALES; SHARE PRICE directly affects MARKETING.

The representation of an experience is a directed graph which we call ***experience map***. In an experience map, concepts are represented as vertexes and causal relationships are represented as directed edges between vertexes. Figure 6.3 is an example of an experience map. From this experience map, we can see how different factors affect each other. For instance, INTERNET SALES is affected by PROMOTION, PRESALE SERVICE, DELIVERY, and ADVERTISING. ADVERTISING is also affected by other three factors: TV, RADIO, and NEWSPAPER.

6.2.2 Experience Elicitation

In this research, experiences are computerized mental models. Thus we can employ cognitive mapping technique to elicit managers' experience. Using the cognitive mapping technique, there are four steps to elicit experience from a manager (Carley & Palmquist 1992; Gnyawali & Tyler 2005).

Step1: Collecting Data
Experience is domain-specific human knowledge. The source data for experience solicitation is generally about a specific topic, e.g., notebook sales. The source

data can be collected via various methods such as post hoc coding of documents, interviews, and direct brainstorming.

Step 2: Identifying Concepts
Concepts can be identified from the source data manually or automatically. Manually extracting concepts can be done by the investigator with the help of the manager. The investigator reads through the texts collected in the previous step and highlights the important key words of interest to the theme topic. The key words are generalized into a set of concepts in light of the manager's domain knowledge. Concept identification can also be done automatically through using text analysis programs, e.g., *Decision Explorer* (as per the web link below)

http://www.banxia.com/dexplore/index.html.

Step 3: Defining Causal Relationships
The identified concepts are presented to the manager. The manager is asked to compare each pair of concepts and examine whether a causal relationship exists between them. Relationships between concepts could be strong or weak. Only strong relationships are kept for experience elicitation. The degree to which causal relationships are strong or weak can be intuitively determined based on the manager's experience. There are also fuzzy cognitive maps where causal relationships are assigned weights (fuzzy values) to indicate the magnitude of causes (Kosko 1986). In this research, we only focus on non-fuzzy causal relationships. For example, For instance, causal relationship $\boldsymbol{r}^c$ (PROMOTION, INTERNET SALES) indicates PROMOTIION is a cause of INTERNET SALES.

Step 4: Validating Experience Maps
Once all the concepts and relationships are extracted and defined, draft experience maps can be produced. Some errors or inappropriateness of representation might be introduced into the draft experience maps. The manager is presented with the experience maps and asked to validate them in terms of business implications. During validation, experience maps are analyzed and might be abandoned or aggregated.

6.2.3 Creating an Experience Base

Managers' experiences are represented as experience maps and stored in an experience base. As experience maps are directed graphs, we need to define vertexes and edges of these graphs. Thus, the experience base consists of directed graphs representing the experience maps.

The only information a vertex of an experience map needs to express is a concept label. In order to avoid ambiguities, we use the corresponding ontology defined in an application to limit the terms used as concept labels. Thus, the concept label of each vertex of an experience map is also a class label in the corresponding class tree. The straightforward adoption of class labels in experience maps results in greater simplicity and clarity for software engineering and end users.

Class label adoption might cause the class tree to grow in a poorly managed manner. Theoretically, an ontology describes the general nature of an application domain: classes and relationships. In a class tree, different classes are well-organized according to the subsumption relationships between them. For example,

MOUNTAIN BIKE, ROAD BIKE and TOURING BIKE are children of BIKE; BIKE (also for COMPONENT, CLOTHING and ACCESSORY) is a child of PRODUCT. This organization of classes is intuitive and easy to understand. However, in order to cater for the complete representation of experience maps, some extra classes might need to be added, such as PRODUCT CATEGORY and PRODUCT SUBCATEGORY. The supplement of these extra classes might damage the well-organized nature of the ontology.

We use adjacency lists to actually store the directed graphs of experience maps. Each vertex of an experience map corresponds to a list of vertexes consisting of this vertex and all other vertexes which have edges **directed to** this vertex. In FACETS, we use a csv file to store the adjacency lists data. This data file is input into the system automatically during the process of experience base creation. The definition file for the experience map in Figure 6.3 is shown in Figure 6.4.

Scholarship

University, Scholarship, Course

Course

Government

Channel, University, Government, Business

Business

Advertisement

Notebook Sales, Channel, Advertisement, Notebook Specification

Graph Card

Notebook Specification, Graph Card, Free Game, HDD Speed, Memory

Memory

Free Game

HDD speed

Fig. 6.4 The Definition File of an Experience Map

In Figure 6.4, the first concept of each list (delimited by blank lines) denotes a vertex; all other concepts in the list are adjacent vertexes from each of which an edge is directed to the first vertex. For example, in the second list, UNIVERSITY is a vertex; SCHOLARSHIP and COURSE are adjacent vertexes to UNIVERSITY. For vertexes to which there are no edges directed from other vertexes, the corresponding adjacency list has only one concept (itself). An experience base might include many experience maps. All the experience maps can be defined in a single csv file.

6.2.4 *Cues*

Confronted with a crime scene, it must be a thrill for detectives if they eventually identify a clue which implicates the true offender. As valuable clues to detectives, similar information and knowledge is also crucial to a business manager in decision making. This kind of information and knowledge might be factors, parameters or data about some business objects. We refer to this kind of information and knowledge hiding in a decision situation as the ***cue*** of the decision situation. Generally speaking, cue can be any information and knowledge which sheds light on the target decision problem of the manager's concern.

Let $\boldsymbol{E}(\boldsymbol{C}, \boldsymbol{R})$ be an experience and $\boldsymbol{c}$ be a concept of $\boldsymbol{E}$ ($\boldsymbol{c} \in \boldsymbol{C}$). We define the cue $\check{\boldsymbol{E}}$ of $\boldsymbol{c}$ in $\boldsymbol{E}$ as follows.

Definition 6.5. Cue ($\check{\boldsymbol{E}}$)

$$\check{\boldsymbol{E}} := (\check{\boldsymbol{C}}, \check{\boldsymbol{R}}),$$

where, $\check{\boldsymbol{C}}$ *is a concept set,*

$\check{\boldsymbol{R}}$ *is a causal relationship set,*

$$\forall c_i \in C,\ c_i \begin{cases} \in \check{C},\ \textit{if}\ \ r^{c}(c_i,c) \text{ or } r^{m}(c_i,c)\ \textit{exists} \\ \notin \check{C}\ \textit{otherwise} \end{cases}$$

and

$$\check{\boldsymbol{R}} := \{\, \boldsymbol{r}^c(\boldsymbol{c}_i, \boldsymbol{c}_j) \mid \boldsymbol{r}^c \in \boldsymbol{R},\ \boldsymbol{c}_i \neq \boldsymbol{c}_j,\ \boldsymbol{c}_i \in \check{\boldsymbol{C}},\ \boldsymbol{c}_j \in \check{\boldsymbol{C}} \,\}.$$

According to Definition 6.5, the cue $\check{\boldsymbol{E}}$ of a concept $\boldsymbol{c}$ in an experience $\boldsymbol{E}$ has two parts: a set of concepts ($\check{\boldsymbol{C}}$) and a set of causal relationships ($\check{\boldsymbol{R}}$) between concepts. The concept $\boldsymbol{c}$ per se is an element of $\boldsymbol{C}$. Furthermore, all other related concepts, from each of which there is a direct or mediated causal relationship to $\boldsymbol{c}$, also belong to $\boldsymbol{C}$. In other words, given a specific experience, the cue of a concept includes all factors in an experience which affect this concept no matter whether it is direct or indirect.

The representation of the cue of a concept is also a directed graph, which we call ***cue maps***. In a cue map, a concept is represented as a vertex and a causal relationship is represented as a directed edge from one vertex to another.

In a decision situation, a concept represents a problem or issue, e.g., sales. The corresponding cue of the concept implies possible factors, information or solutions to the problem. Cues play two roles in the process of situation retrieval.

(1) The cues of all concepts extracted from managers' SA are used to construct data warehouse queries (Chapter 8). This insures the information retrieved from the data warehouse based on the generated queries is valuable to the manager's decision situation.

(2) Cues are used to generate navigation knowledge for situation presentation. Navigation knowledge guides the interaction between the manager and the decision support system (Chapter 9). The experiment results show that information presentation guided by the navigation knowledge is intuitive, and users feel more comfortable with it.

6.2.5 Extracting Cues

Given an experience and a concept in this experience, the algorithm to extract cues is as follows:

Algorithm 6.1. CueExtraction

Input:

(1) Experience (***E***)

(2) A concept (***c***)

Output:

A cue map ($\check{\boldsymbol{E}}$)

Procedure:

Step 1. Find the vertex ***c*** in ***E***.

Step 2. Create an empty adjacency list ***l***.

Step 3. Get all cause concepts of ***c*** in ***E***.

Step 4. Set ***c*** as the first concept of ***l*** and insert all cause concepts into ***l***.

Step 5. Insert ***l*** into $\check{\boldsymbol{E}}$.

Step 6. For each cause concept $\boldsymbol{c}_i$ of ***c***, if the adjacency list of $\boldsymbol{c}_i$ has not been created in $\check{\boldsymbol{E}}$, recursively call this algorithm using ***E*** and $\boldsymbol{c}_i$ as input.

Step 7. End.

CueExtraction is a recursive algorithm. Given a concept, it firstly generates an adjacency list for that concept. Then the algorithm invokes itself to generate adjacency lists for all other concepts in the adjacency list of that concept. This algorithm allows cycles in the experience maps which represent the situation that concepts affect each other. We use an example to illustrate Algorithm 6.1.

An abstract experience is shown in Figure 6.5. Let this experience be $\boldsymbol{E} := (\boldsymbol{C}, \boldsymbol{R})$. In experience $\boldsymbol{E}$, all concepts, except for $\boldsymbol{d}$, are cause concepts; Concepts $\boldsymbol{c}, \boldsymbol{f}$, and $\boldsymbol{d}$ are effect concepts.

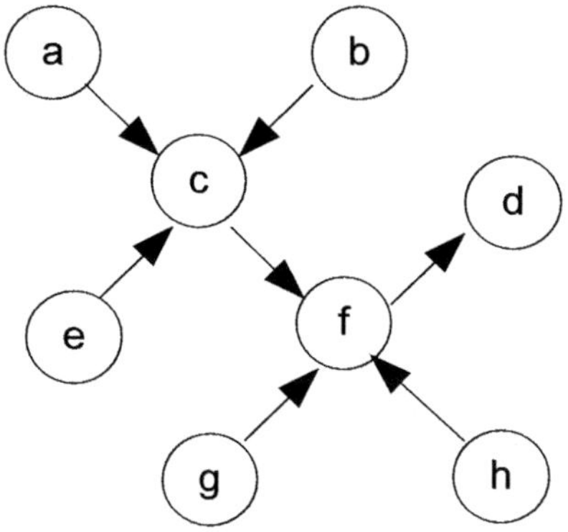

Fig. 6.5 An Abstract Experience Map

The concept set

$$\boldsymbol{C} = \{a, b, c, d, e, f, g, h\}.$$

The causal relationship set

$$\boldsymbol{R} = \{ r^c(a, c), r^c(b, c), r^c(e, c), r^c(c, f), r^c(g, f), r^c(h, f), r^c(f, d)\}.$$

All mediated causal relationships are

$$r^m(a, d), r^m(b, d), r^m(c, d), r^m(e, d), r^m(a, f), r^m(b, f) \text{ and } r^m(e, f).$$

Using Algorithm 6.1, the cues of different concepts can be extracted.

$\check{E}(a) = (\{a\}, \varnothing)$

$\check{E}(b) = (\{b\}, \varnothing)$

$\check{E}(c) = (\{a, b, e\}, \{ r^c(a, c), r^c(b, c), r^c(e, c)\})$

$\check{E}(f) = (\{a, b, c, e, f, g, h\}, \{ r^c(a, c), r^c(b, c), r^c(e, c), r^c(c, f), r^c(g, f), r^c(h, f) \})$

$\check{E}(d) = (\{a, b, c, d, e, f, g, h\}, \{ r^c(a, c), r^c(b, c), r^c(e, c), r^c(c, f), r^c(g, f), r^c(h, f), r^c(f, d)\})$

Cue maps $\check{E}(c)$ and $\check{E}(f)$ are shown in Figure 6.6 (i) and (ii) respectively.

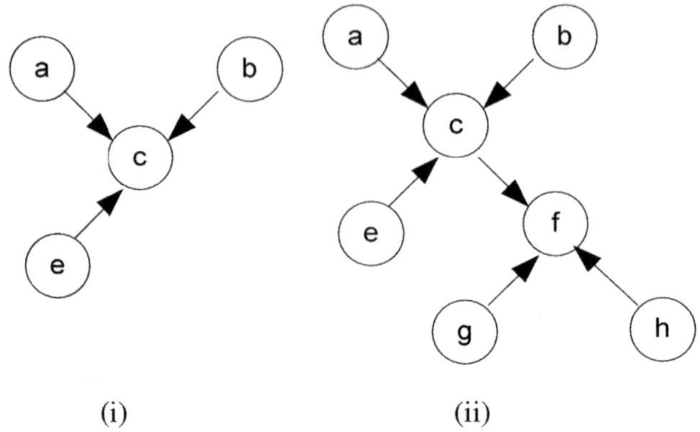

Fig. 6.6 Cue Maps

6.2.6 Knowledge Retrieval

Knowledge retrieval is the process of searching for relevant experience in the experience base. Knowledge retrieval is carried out according to knowledge needs which are extracted from managers' SA descriptions.

Let $\boldsymbol{EB} = \{\boldsymbol{E}_1, \boldsymbol{E}_2, \ldots, \boldsymbol{E}_n\}$ be an experience base, $\boldsymbol{KN} = \{\mathbf{c}_1, \mathbf{c}_2, \ldots, \mathbf{c}_n\}$ be a knowledge need. $\boldsymbol{E}_i$ is an experience; $\boldsymbol{c}_i$ is a local context. We developed algorithm ***KnowledgeRetrieval*** for retrieving relevant experience from ***EB*** according to ***KN***.

Algorithm 6.2. Knowledge Retrieval

Input:

(1) Experience base (***EB***)

(2) A knowledge need (Klein et al.)

Output:

A cue set ***ĚS***

Procedure:

Step 1. Get a concept $\boldsymbol{c}_i$ from ***KN***.

Step 2. In ***EB***, find the most similar concept $\boldsymbol{c}$ to $\boldsymbol{c}_i$ based on class similarity. If the class similarity sim($\boldsymbol{c}$, $\boldsymbol{c}_i$) is greater than a pre-set similarity threshold, go to Step 3. Otherwise, go to Step 4.

Step 3. Call algorithm *CueExtraction* to extract the cue of $\boldsymbol{c}$. Insert this cue into ***ĚS***. Go to Step 4.

Step 4. If all concepts in ***KN*** has been processed, go to Step 5, otherwise go to Step 1.

Step 5. End.

In Step 2, the most similar concept $\boldsymbol{c}$ to $\boldsymbol{c}_i$ is searched through assessing its similarity with all concepts in the experience base. It is possible that $\boldsymbol{c} = \boldsymbol{c}_i$, i.e., $\boldsymbol{c}_i$ is a concept of the experience base ***EB***. If the similarity threshold is set as 1, only when concept $\boldsymbol{c}_i$ exists in ***EB***, the corresponding cue will be extracted. The result of the knowledge retrieval for a knowledge need is a set of cues. This cue set includes all potential cues of the decision situation represented by the knowledge need.

6.2.7 Generating Navigation Knowledge

Navigation knowledge is defined as a special kind of knowledge which acts a mechanism for situation presentation. Navigation knowledge is generated via integrating relevant cues. We developed the algorithm ***Navigator*** for navigation knowledge generation.

Algorithm 6.3. Navigator

Input:

A set of cues $\check{\boldsymbol{E}}\boldsymbol{S} = \{\check{\boldsymbol{E}}_1, \check{\boldsymbol{E}}_2, \ldots, \check{\boldsymbol{E}}_n\}$

Output:

A navigation knowledge: $\boldsymbol{N}$

Procedure:

Step 1. Let $\check{\boldsymbol{E}}_0 \leftarrow \varnothing$ and $\boldsymbol{i} \leftarrow 1$.

Step 2. Get a cue $\check{\boldsymbol{E}}_i$ from $\check{\boldsymbol{E}}\boldsymbol{S}$.

Step 3. Get a causal relationship $\boldsymbol{r}^c(\boldsymbol{c}, \boldsymbol{c}')$ in $\check{\boldsymbol{E}}_i$.

Step 4. If $\boldsymbol{r}^c(\boldsymbol{c}, \boldsymbol{c}') \in \check{\boldsymbol{E}}_0$, go to Step 5. Otherwise,

if $\boldsymbol{c} \in \check{\boldsymbol{E}}_0$, and $\boldsymbol{c}' \notin \check{\boldsymbol{E}}_0$, add a node $\boldsymbol{c}'$ then add a causal relationship $\boldsymbol{r}^c(\boldsymbol{c}, \boldsymbol{c}')$ into $\check{\boldsymbol{E}}_0$.

If $\boldsymbol{c} \notin \check{\boldsymbol{E}}_0$, and $\boldsymbol{c}' \in \check{\boldsymbol{E}}_0$, add a node $\boldsymbol{c}$ then add a causal relationship $\boldsymbol{r}^c(\boldsymbol{c}, \boldsymbol{c}')$ into $\check{\boldsymbol{E}}_0$.

If $\boldsymbol{c} \notin \check{\boldsymbol{E}}_0$, and $\boldsymbol{c}' \notin \check{\boldsymbol{E}}_0$, add two nodes $\boldsymbol{c}$ and $\boldsymbol{c}'$, then add a causal relationship $\boldsymbol{r}^c(\boldsymbol{c}, \boldsymbol{c}')$ into $\check{\boldsymbol{E}}_0$.

If $\boldsymbol{c} \in \check{\boldsymbol{E}}_0$, $\boldsymbol{c}' \in \check{\boldsymbol{E}}_0$, $\boldsymbol{r}^c(\boldsymbol{c}, \boldsymbol{c}') \notin \check{\boldsymbol{E}}_0$, add a causal relationship $\boldsymbol{r}^c(\boldsymbol{c}, \boldsymbol{c}')$ into $\check{\boldsymbol{E}}_0$.

If all causal relationships in $\check{\boldsymbol{E}}_i$ have been processed, go to Step 5, otherwise go to Step 3.

Step 5. If $\boldsymbol{i} < \boldsymbol{n}$, $\boldsymbol{i} \leftarrow i+1$, go to Step 2. Otherwise, go to Step 6.

Step 6. $\boldsymbol{N} \leftarrow \check{\boldsymbol{E}}_0$.

Step 7. End.

As cues are represented as directed graphs (cue maps), the generation of navigation knowledge is a merge process of graphs. Algorithm *Navigator* first creates an empty cue $\check{E}_0$ and then tries to insert all cues in the cue set into $\check{E}_0$ one by one. For each cue to be inserted, *Navigator* gets a causal relationship (edge) of this cue. If either of the two concepts (the two nodes of the edge) is not in $\check{E}_0$, the node and the corresponding causal relationship will be created in $\check{E}_0$. Note that, causal relationships are represented as directed edges. $r^c(c, c')$ and $r^c(c', c)$ are different causal relationships. Hence, if the two concepts already exist in $\check{E}_0$, but the causal relationship between them is $r^c(c', c)$, a new causal relationship $r^c(c, c')$ will be created.

6.3 Summary

Domain knowledge is the fundamental part of the CDDP model. We proposed a set of methods and algorithms to represent domain knowledge for business decision situations. Main points of the proposed methods and algorithms are as follows.

(1) We defined property-share relationships for indirect information acquisition.

(2) We defined class trees and class graphs for ontology representation.

(3) We discussed the method of creating class trees, class graphs and experience base.

(4) We proposed a method to compute class similarities based on the class tree and the class graph.

(5) We defined causal relationships and experiences.

(6) We defined the concept of cue and developed an algorithm for cue extraction.

(7) We developed an algorithm for knowledge retrieval and an algorithm for generating navigation knowledge.

According to the CDDP model, managers describe their SA in the form of natural language. Domain ontology is used to extract semantic information from managers' SA, which is the major concern of Chapter 7.

Chapter 7
Natural Language Processing for Situation Awareness

We employ natural language processing (NLP) techniques to extract semantic information from SA descriptions. According to the CDDP model, managers can describe their SA using a natural language (English) and input SA descriptions into the system. This feature makes it easier for managers to interact with the system and develop their SA for decision making. SA descriptions are comprised of SA sentences. A syntactic parser is used to chunk down SA sentences into words and phrases, and extract syntactic relationships between words or phrases. Based on the results of the syntactic parser, instances, classes, SA triples and local contexts can be extracted or inferred. We refer to the whole process as SA parsing. A general English syntactic parser, Link Grammar Parser, is introduced in Section 7.1. SA parsing is, based on the information types defined in this book, discussed in Section 7.2. Section 7.3 is the overview of the process of SA parsing. The detailed sub-processes of SA parsing are presented in Sections 7.4, 7.5 and 7.6.

7.1 Link Grammar Parser

Link Grammar Parser (Temperley & Sleator 1993) is an English syntactic parser. Given a sentence, this parser generates a syntactic structure called *linkage* consisting of a number of *links*. Each link represents a syntactic relationship between two words. An example of linkage is shown in Figure 7.1. This linkage consists of 12 links. A link is denoted by a dashed line connecting two words. Words are followed by their POS tags. Each link is labeled with a link type. For instance, the link type between *We* and *need* is *Sp*. A *Sp* link connects a subject-noun (*We*) to a finite verb (*need*). Note that the Link Grammar Parser will insert an artificial word at the beginning and end of every sentence before it is parsed, known as the *wall*.

Based on the linkage of a sentence, the Link Grammar Parser also chunks sentences into phrases and produces constituent trees. Figure 7.2 is the corresponding constituent tree for the linkage shown in Figure 7.1. This constituent tree can also be visualized as a tree-like structure shown in Figure 7.3.

Using the Link Grammar Parser, SA sentences can be chunked down into words or phrases. Meanwhile, different syntactic relationships between these words or phrases can be identified. Based on the result of the Link Grammar

L. Niu, J. Lu, and G. Zhang: Cognition-Driven Dec. Supp. for Business Intel., SCI 238, pp. 97–118.
springerlink.com

```
+----------------------------------------Xp----------------------------------------+
|          +-----------------TOo-----------------+                                   |
|          +---------------Os--------------+     |                                   |
|          |   +------------Dsu------------+     |     +-----------Op----------+     |
|          |   |     +----------Am---------+     |     |     +-------Dmc------+     |
+--Wd--+-Sp-+   |     |          +-----AN----+     +--I--+     |      +---AN---+     |
|      |    |   |     |          |           |     |     |     |      |        |     |
LEFT-WALL we need.v a better.a marketing.n strategy.n to improve.v the notebook.n sales.n .
```

Fig. 7.1 A Linkage Produced by Link Grammar Parser

```
(S (NP We)
   (VP need
       (NP a better marketing strategy)
       (S (VP to
              (VP improve
                  (NP the notebook sales)))))
   .)
```

Fig. 7.2 A Constituent Tree Produced by Link Grammar Parser

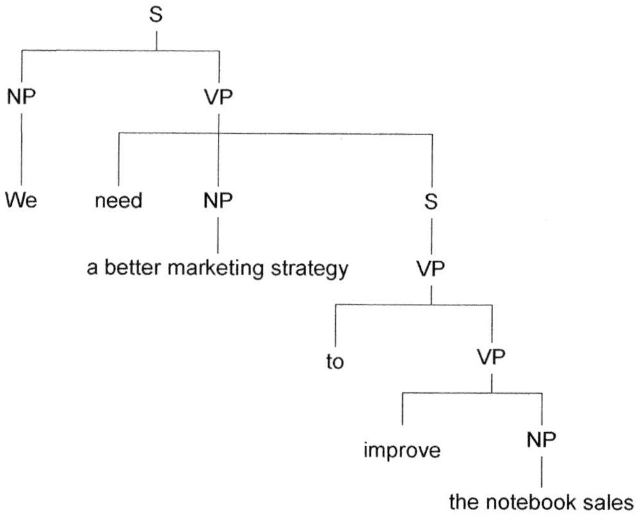

Fig. 7.3 Tree-Like Presentation of a Constituent Tree Produced by Link Grammar Parser

Parser and an ontology, relevant instances and classes of the ontology can be extracted or inferred. The words, phrases, instances and classes are semantically isolated. We refer to them as ***SA tokens***. Thus, if we use a SA token set to collect all SA tokens generated in the course of SA parsing, this SA token set initially only contains the words and phrases output by the parser. Then instances and classes are gradually added into the SA token set. By annotating SA tokens based on the domain ontology, SA triples and local contexts can be produced. In the following sections, we will discuss the detailed process of SA parsing.

7.2 Information Types

From an IS perspective, instances of a class are data representing low level facts. Data always comes with specific data types. Database management systems have a set of pre-defined data types. For example, in SQL Server 2005, there are such data types as *int, money, float, datetime, nchar* and *nvarchar*. In order to expose the semantic information of SA descriptions, data needs to be interpreted as information. Correspondingly, data types also need to be re-explained from a semantic perspective. We use the term ***information type*** to represent the different types of semantic information that SA instances carry.

We defined 19 information types (Table 7.1) in this research, each of which corresponds to a meta class except *general* type (code: 0). Note that this definition

Table 7.1 Information Types in the CDDP model

Code	Information Type	Corresponding Class	Example
1	area	AREA	"20 acres"
2	speed	SPEED	" 60 KM/H"
3	money	CURRENCY	"$26,000.00"
4	temperature	TEMPERATURE	"180 ℃"
5	distance	DISTANCE	"45 kilometers"
6	time interval	TIME INTERVAL	" 10 years"
7	time point	TIME POINT	"2007-6-25"
8	start time	TIME POINT	"8:00am"
9	end time	TIME POINT	"8:00am"
10	length	LENGTH	"20 cm"
11	height	HEIGHT	"20 cm"
12	volume	VOLUME	"45 liter"
13	power	POWER	800 KW
14	weight	WEIGHT	"5 KG"
15	pressure	PRESSURE	1.02 KPa
16	country	COUNTRY	"Australia"
17	state	STATE	"NSW"
18	city	CITY	"Sydney"
0	general	n/a	"Mountain-10"

attempts to cover the most common information types based on our experience; it is by no means an exhaustive list. The defined information types can be roughly divided into two categories: ***numeric group*** and ***literal group***. In Table 7.1, the first 15 information types (Codes 1 to 15) belong to the numeric group; others (16-18 and 0) are in the literal group. Instances of literal types can usually be easily enumerated, such as *company names, product models,* and *colors*. By comparison, it is generally hard to enumerate numeric instances such as *age, income, sales amount and time*.

7.3 The Process of Situation Awareness Parsing

Theoretically, information is supposed to be communicated without semantic uncertainties, other than deliberate ones. Thus, every instance existing in a SA description is intended to belong to a unique class in the ontology from a semantic perspective, although syntactically it is a many-many mapping relationship between instances and classes. Let us look at an example:

Sydney is a beautiful city.

In this SA description, the term *Sydney* is an instance and we can have an inference as follows:

Sydney is the name of a ***place****.*

This inference is based on the intended meaning (semantics) of this SA description. However, without semantic understanding, *Sydney* might refer to totally different things. For example,

Sydney is the name of a ***person****.*

From a syntactic perspective, this inference does not conflict with the original SA description. In this case, uncertainties will be introduced unless definite semantics can be extracted.

The basic process of SA parsing includes four steps shown in Figure 7.4.

Step 1. Syntactic Parsing
The Link Grammar Parser is employed to conduct syntactic parsing on the SA. As discussed in Section 7.1, the Link Grammar Parser can chunk down SA sentences into words and phrases (initial SA tokens); it can also identify the syntactic relationship between words or phrases.

Step 2. Plain Parsing
During plain parsing, meta instances are directly extracted from the initial SA tokens according to simple parsing rules: regular expressions and the lexicon. Meta instances are the instances of meta classes defined in an ontology, such as *July 2007, Mountain-100 Silver, $1200* and *3 weeks*. Regular expressions are patterns for string matching. The lexicon is used to store all possible literal meta instances. The lexicon can be automatically created through importing relevant

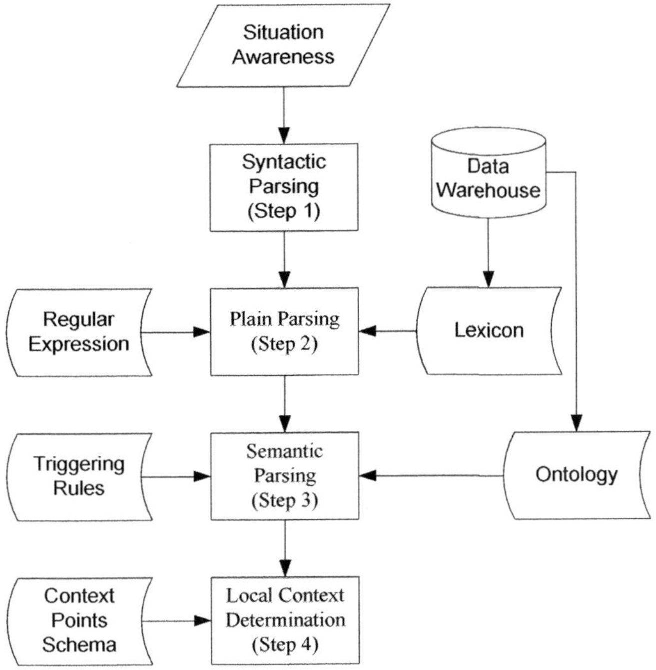

Fig. 7.4 Situation Awareness Parsing

meta instances from a data warehouse. Regular expressions and the lexicon are used to extract numeric and literal meta instances respectively. In plain parsing, provisional SA triples are also generated. There are uncertainties in the provisional SA triples, which can be reduced during semantic parsing.

Step 3. Semantic Parsing
Based on the results of plain parsing and the domain ontology, general and abstract classes are inferred, such as INTERNET SALES, PRODUCT and PRODUCT RELEASE. Compared to meta instances recognized by plain parsing, classes carry richer semantic information which can be used to re-explain (annotate) the recognized meta instances, for example, ***July 2007*** *is a start date of* **INTERNET SALES**. We call this level of SA parsing ***semantic parsing***.

Step 4. Local Context Determination
We proposed a context point schema to determine the local context for a SA sentence based on the results of plain parsing and semantic parsing.

7.4 SA Plain Parsing: Instance Recognition

Plain parsing recognizes meta instances from SA descriptions. Based on the definition of information types, there are two kinds of meta instances to be recognized: *numeric meta instances* and *literal meta instances*.

7.4.1 Numeric Meta Instances

We developed an algorithm (***NumericPlainParser***) for the recognition of numeric meta instances from SA descriptions. In *NumericPlainParser*, numeric meta instances are recognized using pattern matching techniques which are based on regular expressions.

Regular expressions are a very powerful pattern matching technique, originating from automata theory and formal language theory (Wikimedia 2008). In programming, pattern matching is a search process in which a source string is searched in order to extract specific information of interest according to pre-defined rules. Regular expressions act as such rules for pattern matching. A regular expression is a search pattern which consists of a set of text strings. Each text string describes a matching rule. During the search process, substrings will be extracted if corresponding matching rules are met. Some examples of regular expressions defined in *NumericPlainParser* are as follows.

```
"(19|20)[0-9][0-9]-(([1-9])|(0[1-9])|(1[0-2]))-(([1-9])|(0[1-9])|([1-2][0-9])|(3[0-1]))"
```

This regular expression is used to extract *date* information of format YYYY-MM-DD, where YYYY, MM, DD respectively represent a year of four digits, a month of two digits and a day of two digits, e.g., *2008-08-25*. The information type of this kind of date information is *time point.*

```
"(([1-9])|(0[1-9])|([1-2][0-9])|(3[0-1]))-(([1-9])|(0[1-9])|(1[0-2]))-(19|20)[0-9][0-9]"
```

This regular expression is used to extract date information of format DD-MM-YYYY, e.g., *20-06-1974*. The information type of this kind of date information is *time point.*

```
"(([1-9])|(0[1-9])|([1-2][0-9])|(3[0-1]))(,|\\s{1,})
(January|February|March|April|May
|June|July|August|September|October|November|December)(,|\\s{1,})(19|20)[0-9]
[0-9]"
```

This regular expression is used to extract date information of the format YYYY-Month-DD, where Month represents a month word, e.g., *2005-July-11*. The information type of this kind of date information is *time point.*

In *NumericPlainParser*, we coded 29 regular expressions for date extraction. Thus, managers can use up to 29 kinds of date format to describe their SA.

Based on information of *time point* type, date information of the *start time* type and *end time* type can also be extracted via checking preposition words. For instance,

```
"from\s{1,}(19|20)[0-9][0-9]-(([1-9])|(0[1-9])|(1[0-2]))-(([1-9])|(0[1-9])|([1-2][0-
9])|(3[0-1]))(!|\(|\)|\{|\}|\[|\]|\|\:|;|"|,|\.|\?|\s{1,})"
```

This regular expression is used to extract date information which represents a start date, such as (*from*) *2003-1-12*. The information type of this kind of date information is *start time*.

```
"to\s{1,}(19|20)[0-9][0-9]-(([1-9])|(0[1-9])|(1[0-2]))-(([1-9])|(0[1-9])|([1-2][0-
])|(3[0-1])) (!|\(|\)|\{|\}|\[|\]|\|\:|;|"|,|\.|\?|\s{1,})"
```

This regular expression is used to extract date information which represents an end date, such as (*to*) *2003-2-10*. The information type of this kind of date information is *end time*.

Algorithm *NumericPlainParser* can dynamically construct regular expressions during run-time to extract *non-date* numeric information. Some examples are as follows.

```
"\s{1,}(\d{1}\d{1}\s{0,}[m|M]{0,1})\s{0,}\s{0,}days(!|\(|\)|\{|\}|\[|\]|\|\:|;|"|,|\.|\?|
\s{1,})"
```

This regular expression is used to extract time interval information of the format *DD days,* e.g., 20 days. The information type of this kind of date information is *time interval.*

```
"\s{1,}( \d{1}\d{1}\d{1}\d{1}\s{0,}[m|M]{0,1})\s{0,}\s{0,}cm(!|\(|\)|\{|\}|\[|\]|\|\:|;
|"|,|\.|\?|\s{1,})"
```

This regular expression is used to extract distance information of format *DDDD cm*, e.g., 1512 cm. The information type of this kind of date information is *distance*.

```
"\s{1,}( \d{1}\d{1}\d{1}\s{0,}[m|M]{0,1})\s{0,}\s{0,}meters(!|\(|\)|\{|\}|\[|\]|\|\:|;|"|,|
\.|\?|\s{1,})"
```

This regular expression is used to extract distance information, e.g., 115 meters. The information type of this kind of date information is *distance*.

As can be seen from the above examples, the length of the digits in regular expressions is dynamic in order to match numbers of different lengths in SA tokens. *NumericPlainParser* permits users to set the maximum length of digits according to their specific applications.

7.4.2 Literal Meta Instances

We developed an algorithm (***LiteralPlainParser***) for the recognition of literal meta instances from SA. Generally speaking, literal instances are enumerable information. Thus, a domain-specific lexicon can be pre-created to cover all possible literal meta instances. As the total number of possible meta instances in an application ranges quite widely from a few to hundreds, it might be very labor-intensive and time-consuming to define the lexicon. In this research, we created mapping relationships between an ontology and a data warehouse system. The domain-specific lexicon can be easily populated through importing instances from the data warehouse.

An item (entry) in the lexicon contains four kinds of information: the value of a meta instance, the corresponding property, the corresponding class and information type. The information type for literal meta instances is *general* (code: 0). The following is an example of lexicon item.

Lexicon item: *Mountain-100 Silver/EnglishProductName/Mountain Bike/0*

where,

Instance value: *Mountain-100 Silver*

Property: *EnglishProductName*

Class: *Mountain Bike*

Information Type: *0*

Excepting literal meta instances, the lexicon also contains information about numeric meta instances. As numeric meta instances are generally non-numerable, for each type of numeric meta instance there is only one corresponding lexicon item with the instance value being empty. For example, *null/Price/Mountain Bike/3* is a lexicon item, where the information type of *Price* is *money*.

LiteralPlainParser uses the lexicon to search for literal meta instances existing in SA descriptions. The searching for literal meta instances in the lexicon is a straightforward boolean matching process. If a match is found between a SA token and the *instance value* of an item in the lexicon, this lexicon item is exported as a SA triple. The corresponding SA triple is created by setting its context, view and wording as the class, property and value of the lexicon item respectively.

When a numeric meta instance is extracted from a SA description, the lexicon will also be searched for matched items with this numeric meta instance. The matching rule is that both the lexicon item and the numeric meta instance have the same information type. If a match is found, this lexicon item is exported as a SA triple. The corresponding SA triple is created by setting its context and view as the class and property of the lexicon item respectively. The numeric meta instance goes into the wording part of the SA triple.

It is possible that multiple matches for the same numeric or literal meta instance are found in the lexicon. For example, the following lexicon items will result in two matches for *Sydney* (a literal meta instance) and two matches for *$1299.00* (a numeric meta instance):

Sydney/City/Internet Sales/0

Sydney/Surname/Customer/0

null/Retail Price/Mountain Bike/3

null/Cost/Mountain Bike/3

If multiple lexicon items are found, all the matched lexicon items will be exported as SA triples. In this case, uncertainties of SA triples will be introduced. These uncertainties will be reduced or eliminated during the process of semantic parsing.

7.4.3 Reference Properties

In plain parsing, SA tokens are also compared to the *properties* of lexicon items. If a SA token is matched with a lexicon item, (i.e., the token is equal to the property of this lexicon item), the property of the lexicon item (the SA token) will be output. This kind of property will be used to reduce the uncertainties of SA triples. We refer to these properties as ***reference properties***.

For example, suppose a SA token is *retail price.* During the search of the lexicon, this SA token will be matched with the following lexicon item:

null/Retail Price/Mountain Bike/3.

The property *Retail Price* will be output as a reference property.

7.5 SA Semantic Parsing: Class Inferring

Semantic parsing is the process of inferring classes. We developed an algorithm (***SemanticParser***) for semantic parsing. The inputs of *SemanticParser* include the ontology, the SA tokens, the lexicon, the SA triples produced in plain parsing, and the linkage generated by the Link Grammar Parser. *SemanticParser* infers ontology classes (both general and abstract classes) from SA and uses these classes to reduce the uncertainties of SA triples.

Algorithm *SemanticParser* infers classes based on ***class triggers***. A class trigger is associated with a specific class. The presence of a class trigger in a SA sentence and/or SA tokens might result in the presence of this class. However, a class trigger does not guarantee that a class will be successfully inferred, unless pre-defined rules are met at the same time. Examples of class trigger for PRODUCT RELEASE are as follows.

{ *product, release* }

{ *product release* }

{ *product, announcement* }

As can be seen from the above examples, a class trigger is a term set consisting of a number of terms, each of which can be a class, a word or a phrase. A class can be triggered by any of its valid class triggers under a triggering rule.

7.5.1 Class Trigger Construction

Let ***C*** be a class in a class tree. We proposed four methods to construct class triggers for class ***C***.

(1) Taxonomy-Property-Based Method

Every instance of the taxonomy property of ***C*** is a complete class trigger. This method is only applicable to general classes. For example, English Product Name is the taxonomy property of class MOUNTAIN BIKE. Some valid class triggers of MOUNTAIN BIKE are as follows:

{*Mountain-100 Silver*}, {*Mountain-80 Silver*}, {*Mountain-100 Black*}.

The element of each class trigger is a meta instance of English Product Name.

(2) Child-Class-Based Method

Every child class of ***C*** in the class tree is a complete class trigger of ***C***. This method is only applicable to general classes. For instance, classes BIKE, COMPONENT, CLOTHING and ACCESSORY are child classes of PRODUCT. Thus, the following are valid class triggers of PRODUCT:

{*bike*}, {*component*}, {*clothing*} {*accessory*}.

(3) Synonym-Based Method

Every synonym of ***C*** is a complete class trigger of ***C***. Class synonyms are defined during ontology development. For example, the synonyms of CUSTOMER are defined as follows:

customers, patron, patrons, buyer, buyers, client, clients, shopper, shoppers.

Thus, each of these synonyms is a class trigger of CUSTOMER.

(4) Semantic-Extension-Based Method

Class triggers constructed using the previous three methods consist of only one element: a meta instance, a child class, or a synonym of the target class. The semantic-extension-based method extends the semantic information of the target class through partial replacement, that is, only part of the class label is replaced by a term each time. A term can be a word or a phrase. Therefore, the semantics extension method only applies to multi-word classes and the constructed class triggers consist of two or more terms. In comparison, the previous three methods can be used both for single-word classes as well as multi-word classes and the constructed trigger classes consist of a single term.

We developed the algorithm ***SemanticExtension*** to implement this method.

Algorithm 7.1. SemanticExtension

Input:

A class (***C***)

Output:

A set of class triggers for ***C***

Procedure:

Step 1. Determine the head noun of the class label of C.

Step 2. Within the class label of ***C***, search for the longest term ***t*** (a sub-term) which is also a class. If ***t*** exists, create an empty trigger member set and insert ***t*** into this set. If ***t*** is the head noun, tag this set as head trigger member set. In the class label of ***C***, replace ***t*** with a dummy string, e.g. '%%%%'. Repeat Step 2, until no more classes can be extracted.

Step 3. Get a word ***w*** from the reminder of the class label of ***C*** (all classes have been taken out at this stage).

Step 4. Create an empty trigger member set and insert all synonyms of ***w*** into this set. If ***w*** is a noun and it has a corresponding verb form, insert the verb and all synonyms of this verb into the trigger member set. If ***w*** is the head noun, tag this set as head trigger member set. If all words in ***C*** have been processed, go to Step 5, otherwise go to Step 3.

Step 5. Collect a term from each of the trigger member sets which have been created and construct a class trigger. Tag the term taken from the head trigger member set as the head of this class trigger. Repeat this step until all combinations are generated.

Step 6. End.

The label of a multi-word class might include other classes which have shorter labels. For example, in class MARKETING STRATEGY MANAGEMENT, both MARKETING and STRATEGY are also classes. As each class in the ontology has its own class triggers, it is unnecessary for a multi-word class to import the class triggers of its subclasses. Therefore, in Step 2 of algorithm

SemanticExtension, sub-terms are not further extended if they are also classes; they are simply treated as a term.

In Step 4, *SemanticExtension* uses a thesaurus to handle non-class words. For a multi-word class, although its class label as a whole is usually a noun phrase, each individual word of the class label might be of different types of POS. For each non-class word, regardless of the type, *SemanticExtension* firstly gets its all synonyms from the thesaurus. If this word is a noun and it has a corresponding verb form, the verb and all the synonyms of the verb will be retrieved from the thesaurus. For example, in PRODUCT RELEASE, *release* is a non-class word and also a noun. The verb form of *release* is also *release*. Thus, the synonyms of *release* (noun) together with the synonyms of *release* (verb) are retrieved from the thesaurus. The trigger member set for *release* is as follows:

{*release, announcement, releases, releasing, released, announces, announcing, announced*}.

Once all trigger member sets are created, class triggers can be constructed. A class trigger is composed through taking a member out of each trigger member set. The final class triggers will be all possible combinations based on the trigger member sets.

If a trigger member set is generated based on the head noun of the class label of ***C***, this set will be tagged as the head trigger member set. Accordingly, the trigger member taken from this set will be tagged as the head of a trigger.

Using *SemanticExtension*, class triggers for PRODUCT RELEASE are generated as follows:

{*product*, ***release***}, {*product*, ***announcement***}, {*product, releases*}, {*product, releasing*}, {*product, released*}, {*product, announces*}, {*product, announcing*}, {*product, announced*}.

Note that, if a class trigger has a head noun, the head noun is printed in bold font in the above example. In real systems, there will need to define extra data structure to indicate the head noun information of class triggers.

7.5.2 *Triggering Rules*

The presence of a class trigger is likely to trigger a class, but does not guarantee. Whether or not a class can be triggered is also subject to triggering rules. We have discussed four methods for constructing class triggers. Accordingly, there are four types of class triggers. The class triggers of the first three types contain only a single element. The fourth type of class triggers contains multiple elements. Note that an element in a class trigger can be a word or a phrase. There are different triggering rules for single-element class triggers and multi-element class triggers.

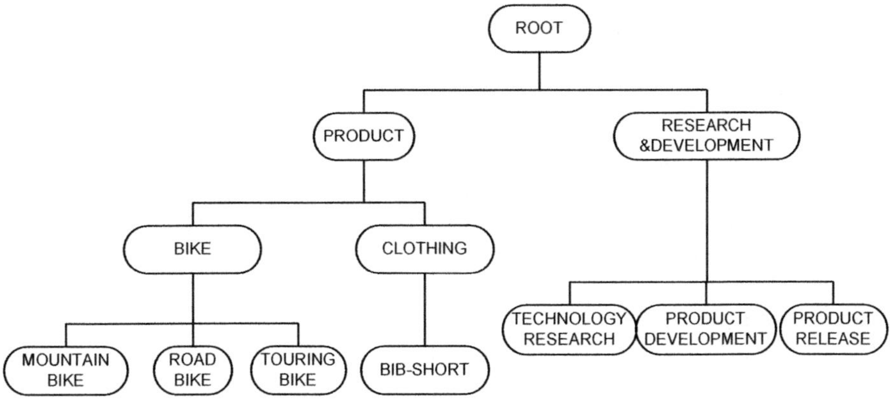

Fig. 7.5 An Example of Ontology

- Single-Element Class Triggers

The triggering rule for single-element class triggers is as follows.

Triggering Rule 7.1

The element of a class trigger can be matched with one of the SA tokens.

This triggering rule is very simple: as long as a class trigger can be found among the SA tokens, the corresponding class can be inferred. Let us look at some examples.

Suppose an ontology is shown in Figure 7.5 and a SA sentence is as follows:

We released Mountain-100 Silver in Australia in 2005.

Algorithm *LiteralPlainParser* can generate a SA triple from this SA sentence:

(MOUNTAIN BIKE, English Product Name, *Mountain-100 Silver*).

The SA token *Mountain-100 Silver*, is a meta instance of the taxonomy property of MOUNTAIN BIKE. According to the taxonomy-property-based method for class trigger construction, {*Mountain-100 Silver*} is a class trigger of MOUNTAIN BIKE and it is also a single-element class trigger. Thus, class MOUNTAIN BIKE can be triggered (inferred) by SA token *Mountain-100 Silver*, i.e.,

Mountain-100 Silver → MOUNTAIN BIKE.

In Figure 7.5, class MOUNTAIN BIKE is a child class of BIKE; BIKE is a child class of PRODUCT. According to the child class based method, {*mountain bike*} is a class trigger of BIKE; {*bike*} is a class trigger of PRODUCT. Thus, classes BIKE and PRODUCT can also be inferred via

MOUNTAIN BIKE → BIKE,

and *BIKE → PRODUCT.*

- Multi-Element Class Triggers

Multi-element class triggers have two basic varieties and they correspond to different triggering rules.

(a) Noun-Noun Class Triggers
This type of class triggers consist of two elements, both of which are nouns, for example, {marketing, strategy} and {internet, sales}. The corresponding triggering rule is as follows.

Triggering Rule 7.2
The two nouns of a class trigger appear in a noun phrase which has the same head noun as the class trigger.

We use an example to illustrate this triggering rule.

Suppose a SA sentence is as follows:

The release of Mountain-100 Silver increased our market share by 45%.

The triggering process is as follows:

Step 1. Triggering Rule 7.1:	*Mountain-100 Silver → MOUNTAIN BIKE*
Step 2. Triggering Rule 7.1:	*MOUNTAIN BIKE → BIKE*
Step 3. Triggering Rule 7.1:	*BIKE → PRODUCT*
Step 4. Triggering Rule 7.2:	*release, PRODUCT → PRODUCT RELEASE*

The first three steps are based on Triggering Rule 7.1. In Step 4, PRODUCT is not actually present in the SA sentence. However, looking back at the process of how PRODUCT is triggered, algorithm *SemanticParser* will replace *Mountain-100 Silver* in the SA sentence with PRODUCT. This replacement makes Triggering Rule 7.2 is met: (1) *release of product* will be output as a noun phrase by the Link Grammar Parser, (2) *release* is the head noun of this noun phrase and the class trigger {*product, **release***}. Therefore, class PRODUCT RELEASE will be successfully triggered by class trigger {*product, **release***}.

(b) Verb-Noun Class Triggers
This type of class triggers consist of two elements: a verb and a noun, for example, {*release, product*} and {*improve, sales*}. There are two corresponding triggering rules as follows.

Triggering Rule 7.3
The noun is the direct object of the verb.

Triggering Rule 7.4
The noun is the subject and the verb is a passive participle.

We use an example to illustrate Triggering Rule 7.3.

Suppose a SA sentence is as follows:

We released Mountain-100 Silver in Australia in 2005.

The triggering process is as follows:

Step 1. Triggering Rule 7.1:	*Mountain-100 Silver → MOUNTAIN BIKE*
Step 2. Triggering Rule 7.1:	*MOUNTAIN BIKE → BIKE*
Step 3. Triggering Rule 7.1:	*BIKE → PRODUCT*
Step 4. Triggering Rule 7.3:	*released, PRODUCT → PRODUCT RELEASE*

In Step 4, *Mountain-100 Silver* is replaced by PRODUCT, such that PRODUCT is the direct object of *released.* Thus, Triggering Rule 7.3 is met and class PRODUCT RELEASE is triggered.

The triggering process for Triggering Rule 7.4 is similar to Triggering Rule 7.3. Based on this rule, class PRODUCT RELEASE can be triggered from the following SA sentence.

Mountain-100 Silver was released in Australia in 2005.

The presence of a class trigger will not trigger the class if the corresponding triggering rule is not met. Look at the following two SA sentences,

Another marketing campaign was ***released*** *after the sales of* ***Mountain-100 Silver*** *went down.*

Occasionally, ***Mountain-100 Silver releases*** *oil when climbing steep ramps.*

As none of triggering rules can be met, class PRODUCT RELEASE cannot be triggered, although the class triggers {released, product} and {product, releases} are present.

The single-element and two-element class triggers correspond to single-word and two-word classes respectively. In an ontology, one-word classes are generally the most common type. Of multi-word classes, two-word classes are the most common type. There are very few classes with the number of words over three. For example, the maximum length (the number of words in class labels) in the Enterprise Ontology (Uschold et al. 1998) is three. There are 62 single-word classes (63%), 34 two-word classes (34%) and 3 three-word classes (3%). Of multi-word classes, 92% are two-word classes and only 8% are three-word classes. In the ontology developed in the application case study I, there are 69 single-word classes (63%), 36 two-word classes (33%) and 5 three-word classes (less than 5%). Therefore, the four triggering rules which have been discussed can deal with about 96% cases of class triggering (in both ontologies). For class triggers whose lengths are three or longer, the corresponding triggering rules become more complex. For example, a class trigger might consist of a verb and a number of nouns. One corresponding triggering rule is as follows:

All nouns appear in a noun phrase and one of them is the head of this phrase. This noun phrase is the direct object of the verb.

7.5.3 Reducing Uncertainties of SA Triples

In plain parsing, uncertainties of SA triples will be generated if multiple lexicon items are matched with a SA token. All the matched lexicon items will be exported as provisional SA triples. However, only of one of the exported SA triples is appropriate in terms the semantics of the SA sentence. Hence, within the uncertain SA triples, it is necessary to determine the rightful one which is best matched with the original SA token. This process is based on reference properties and inferred classes. We will discuss two kinds of uncertainties of SA triples: ***view uncertainties*** and ***context uncertainties***.

- View Uncertainties

The view of a SA triple corresponds to a property of a class or the property of a lexicon item. A view uncertainty between two SA triples is the difference between their views. Let $\boldsymbol{T_1}$ and $\boldsymbol{T_2}$ be two SA triples. The view uncertainty between $\boldsymbol{T_1}$ and $\boldsymbol{T_2}$ is shown as follows.

$$\boldsymbol{T_1} = (\boldsymbol{c},\ \boldsymbol{v_1},\ \boldsymbol{w}),$$

$$\boldsymbol{T_2} = (\boldsymbol{c},\ \boldsymbol{v_2},\ \boldsymbol{w}).$$

SA triples $\boldsymbol{T_1}$ and $\boldsymbol{T_2}$ have the same context and wording, but different view. View uncertainties can be reduced by applying reference properties generated during plain parsing. Let $\boldsymbol{p}$ be a reference property. The rightful SA triple is the one with its view equal to $\boldsymbol{p}$. Suppose $\boldsymbol{v_2} = \boldsymbol{p}$. Thus, $\boldsymbol{T_2}$ will be the rightful SA triple.

- Context Uncertainties

The context uncertainty between two SA triples is the difference between their contexts. Let $\boldsymbol{T_1}$ and $\boldsymbol{T_2}$ be SA triples. The view uncertainty is shown as follows.

$$\boldsymbol{T_1} = (\boldsymbol{c_1},\ \boldsymbol{v},\ \boldsymbol{w}),$$

$$\boldsymbol{T_2} = (\boldsymbol{c_2},\ \boldsymbol{v},\ \boldsymbol{w}).$$

Context uncertainties can be reduced by applying inferred classes which are generated during semantic parsing. These inferred classes are referred to as ***reference classes***. Let $\boldsymbol{c}$ be a reference class. The rightful SA triple is the one with its context equal to $\boldsymbol{c}$. Suppose $\boldsymbol{c_1} = \boldsymbol{c}$. Thus, $\boldsymbol{T_1}$ will be the rightful SA triple.

There are hybrid uncertainties which are the combination of view uncertainties and context uncertainties, meaning that differences exist in both the contexts and views. Hybrid uncertainties can be reduced by respectively applying reference classes and reference properties.

- Syntactic Distance

The above two methods to reduce SA triple uncertainties directly utilize either a reference property or reference class. However, it is not that straightforward in some cases. For example, if two reference properties/classes are applicable to a view/context uncertainty at the same time, there will be a problem of how to choose the rightful one. The solution to this problem is using ***syntactic distance***. A syntactic distance between two words/phrases in a SA sentence is defined as their shortest path length in the corresponding constituent tree of this SA sentence. Thus the syntactic distance between two SA tokens is evaluated based on the constituent tree of the SA sentence.

According to the processes of SA parsing, a SA token can always be tracked back to the original words in the SA sentence, no matter this SA token is an instance, property or class. In order to compute the syntactic distance between a SA token and a reference property/class, they need to be tracked back to the original word/words in the SA sentence. We use the following example to illustrate this process.

Let $\boldsymbol{S}$ be a SA sentence, $\boldsymbol{T}$ be a SA token and $\boldsymbol{R_1}$ and $\boldsymbol{R_2}$ be reference properties/classes. Suppose there are the following inferring relationships:

$\boldsymbol{S} = (\boldsymbol{w_1}, \boldsymbol{w_2}, \ldots, \boldsymbol{w_{10}})$, $\boldsymbol{w_i}$ is a word.

$\boldsymbol{w_1} \rightarrow \boldsymbol{T} \rightarrow$ multiple SA triples

$(\boldsymbol{w_2}, \boldsymbol{w_4}, \boldsymbol{w_6}) \rightarrow \ldots \rightarrow \boldsymbol{R_1}$

$(\boldsymbol{w_3}, \boldsymbol{w_5}, \boldsymbol{w_7}) \rightarrow \ldots \rightarrow \boldsymbol{R_2}$

The SA sentence $\boldsymbol{S}$ consists of ten words. SA token $\boldsymbol{T}$ is inferred from $\boldsymbol{w_1}$. $\boldsymbol{T}$ results in multiple SA triples (uncertainty). Reference property/class $\boldsymbol{R_1}$ is inferred from some classes which are originally inferred from words $\boldsymbol{w_2}$, $\boldsymbol{w_4}$, $\boldsymbol{w_6}$. $\boldsymbol{R_2}$ is inferred from some classes which are originally inferred from words $\boldsymbol{w_3}$, $\boldsymbol{w_5}$, $\boldsymbol{w_7}$.

The syntactic distance ($\boldsymbol{sd}$) between $\boldsymbol{R_1}$ and $\boldsymbol{T}$ can be computed using equation

$$\boldsymbol{sd}(\boldsymbol{R_1}, \boldsymbol{T}) = \frac{1}{3}\left(\boldsymbol{len}(\boldsymbol{w_1}, \boldsymbol{w_2}) + \boldsymbol{len}(\boldsymbol{w_1}, \boldsymbol{w_4}) + \boldsymbol{len}(\boldsymbol{w_1}, \boldsymbol{w_6})\right),$$

where, $\boldsymbol{len}(\boldsymbol{w_i}, \boldsymbol{w_j})$ is the shortest path length from node $\boldsymbol{w_i}$ to node $\boldsymbol{w_j}$ $\boldsymbol{b}$ in the constituent tree.

The syntactic distance ($\boldsymbol{sd}$) between $\boldsymbol{R_2}$ and $\boldsymbol{T}$ can be computed using equation

$$\boldsymbol{sd}(\boldsymbol{R_2}, \boldsymbol{T}) = \frac{1}{3}\left(\boldsymbol{len}(\boldsymbol{w_1}, \boldsymbol{w_3}) + \boldsymbol{len}(\boldsymbol{w_1}, \boldsymbol{w_5}) + \boldsymbol{len}(\boldsymbol{w_1}, \boldsymbol{w_7})\right).$$

The reference property/class syntactically closer (smaller $\boldsymbol{sd}$) to $\boldsymbol{T}$ will be selected. Thus, if multiple reference properties/classes are found, the closest one to the target SA token will be used to reduce the uncertainty of SA triples.

Uncertainties of SA triples cannot be eliminated completely in some cases, due to the complexity, dynamics and uncertainties of natural language expressions. For example, if the SA description per se is vague, it will be extremely difficult, even impossible, to remove the uncertainties of SA triples. In this situation, managers will be questioned to make the final choice of rightful SA triples or rephrase their SA descriptions.

7.6 Local Context Determination

A local context of a SA sentence represents the common background where the majority of SA triples are communicated. All classes which are inferred during semantic parsing are alternatives for the local context.

We developed the algorithm (***LocalContextDetermination***) to determine local contexts for SA sentences. In this algorithm, we use a point system to select the local context from these alternatives.

In this point system, the points of an alternative context reflect its competitiveness to be selected as the local context. We refer to theses points as its ***context points*** (***CP***). We defined three metrics to evaluate the context points of an alternative context from three aspects: ***context coverage point*** (***CCP***), ***context position point*** (***CPP***) and ***inverse context specificity point*** (***ICSP***).

7.6.1 Context Position Points

The *CPP* points of an alternative context are related to its position in the SA sentence. A complete English sentence is made up of a number of POS, each of which has a specific position in the sentence. Different positions imply different significances of information expression. For example, the subject of a sentence is usually more significant than an adverb. We assume that an alternative context falling in a more important position in a sentence will be more competitive to be the local context of this sentence. We defined a point schema in Table 7.2 to assess the *CPP* points for alternative contexts.

The point schema divides a complete sentence into a number of POS. Different POS are positioned differently within the sentence. The points that a POS in a specific position can earn are specified accordingly in the schema. For example, the first POS (at Position 1) represents a pre-subject adjective modifier which is assigned 2 *CPP* points. If a term of a SA sentence is at this position, it will earn 2 *CPP* points. The position of a term in a sentence is determined by algorithm *LocalContextDetermination* based on the links produced by the Link Grammar Parser.

Example. Sentence: *Knowledgeable university graduates in Australia have better employment opportunities in industry.*

Table 7.2 Point Schemas for Context Position and Context Coverage

Position	Part of Speech	CPP	CCP	Example
1.	pre-subject adjective modifier	2	1	*knowledgeable*
2.	pre-subject noun modifier	3	1	*university*
3.	head noun (subject)	5	1	*graduates*
4.	post-subject modifier (prepositional phrase/relative clause)	3	1	*in Australia*
5.	head verb (predicator)	4	1	*have*
6.	verb modifier	1	1	-
7.	pre-object adjective modifier	1	1	*better*
8.	pre-object noun modifier	2	1	*employment*
9.	direct/indirect object	3	1	*opportunities*
10.	post-noun modifier (prepositional phrase/relative clause)	2	1	*in industry*

If a term falls in the first two POS, it will earn another 3 *CPP* points. The final *CPP* points of an alternative context are the sum of points it earns across the whole sentence. Let us look at an example.

SA Sentence: *We released Mountain-100 Silver in Australia in 2005*
Alternative Contexts: PRODUCT, PRODUCT RELEASE

The inferring process for class PRODUCT is

Mountain-100 Silver → MOUNTAIN BIKE → BIKE → PRODUCTThus, PRODUCT can be tracked back to *Mountain-100 Silver* falling in Position 8 (direct/indirect object). Thus, the PRODUCT earns 3 *CPP* points. Similarly, PRODUCT RELEASE falls in Position 5 (head verb (predicator)) and Position 8 (direct/indirect object). Thus PRODUCT RELEASE has 7 *CPP* (4 + 3) points.

The points assigned to different positions in Table 7.2 are on scale of 1 to 5, which were determined by language experts in this research. We have conducted experiments to testify the validity of this schema (Chapter 10). Another method to create the point schema is through machine learning. We are currently working on the neutral network technique to automatically create the point schema.

7.6.2 Context Coverage Points

The *CCP* points of an alternative context are related to its coverage across the SA sentence. We assume that an alternative context with wider coverage in a sentence is more competitive to be the local context of this sentence. A complete English sentence is made up of a number of POS, each of which has a specific position in the sentence. Each POS in the sentence is assigned one *CCP* point (Table 7.2). The more POS an alternative context covers, the more points it can earn. For example,

A SA Sentence:

We released Mountain-100 Silver in Australia in 2005.

Alternative Contexts:

PRODUCT, PRODUCT RELEASE

According to *CCP* schema, PRODUCT covers only one POS (Position 8). It earns 1 *CCP* point. PRODUCT RELEASE earns 2 *CCP* points.

7.6.3 Inverse Context Specificity Points

The *ICSP* points of an alternative context are related to its relative position in a class tree. We use term ***class specificity*** to describe the degree to which a class conveys detailed information. We assume that an alternative context with lower specificity (more abstract) in the ontology is more competitive to be the local context of this sentence. Thus the *ICSP* points of a class are inversely proportional to its class specificity.

Classes in different positions in a class tree correspond to different class specificities. Classes become more specific as one moves toward the leaf nodes in a class tree. The class specificity of a class is determined by its distance to the nearest leaf node in the class tree.

Let ***c*** be a class and ***DTL*** be the proportion of *ICSP* points attributed to the position of ***c*** in the class tree. We use the following equation to compute the *DTL* of ***c***:

$$DTL = d + 1,$$

where, ***d*** is the distance from class ***c*** to the nearest leaf node.

For the example in Section 7.6.2, the *DTL* values for PRODUCT and PRODUCT RELEASE can be calculated as follows, based on the ontology used in application case 1, Chapter 11.

DT L (PRODUCT) = 3

DTL (PRODUCT RELEASE) = 1.

The specificity of a class is also related to the semantic parsing level. As discussed in Section 7.5, semantic parsing has two varieties: inferring general classes and inferring abstract classes. A general class is inferred by the meta instances of its taxonomy property. We refer to the semantic parsing to infer general classes as Level 1 parsing. On the basis of general classes, abstract classes can be inferred, which we call Level 2 parsing. Similarly, classes inferred at Level 2 parsing lead to Level 3 classes, and so on. There are situations where an abstract class is inferred from a number of classes which are not at the same level in the class tree, i.e., the trigger elements are distributed in different positions in the class tree. In this case, we assess the parsing level of this class by averaging all the corresponding lower parsing levels. We use ***class abstract level*** (Microsoft 2007b) to refer to the proportion of *ICSP* points attributed to this kind of class specificity.

The *ICSP* of a class are computed by the following equation

$$ICSP = DTL + CAL.$$

For the example in Section 7.6.2, the *CAL* values of classes PRODUCT and PRODUCT RELEASE calculated by algorithm *LocalContextDetermination* are as follows.

CAL (PRODUCT) = 3
CAL (PRODUCT RELEASE) = 4

Thus, the ICSP values are as follows.

ICSP (PRODUCT) = 3+3 = 6
ICSP (PRODUCT RELEASE) = 1+ 4 = 5.

7.6.4 Local Contexts

The *CP* points of an alternative context are the sum of its *CPP* points, *CCP* points and *ICSP* points, i.e.,

$$CP = CPP + CCP + ICSP.$$

Based on this equation, the local context of a SA sentence can be determined as the alternative context with the highest *CP* points.

For the example in Section 7.6.2, the *CP* values of classes PRODUCT and PRODUCT RELEASE calculated by algorithm *LocalContextDetermination* are as follows.

CP (PRODUCT) = 3 + 1 + 6 = 10
CP (PRODUCT RELEASE) = 7 + 2 + 5 = 14

The local context of SA sentence

We released Mountain-100 Silver in Australia in 2005

is PRODUCT RELEASE.

7.7 Summary

Based on NLP technologies, we proposed a set of methods and algorithms to parse managers' SA. The main points of the proposed methods and algorithms are as follows.

(1) We proposed the concept of information type and defined 19 information types. Compared to data types, information types expose richer semantic information of data, which form one of the important bases for parsing managers' SA.

(2) We developed algorithms to recognize meta instances from SA, which we call plain parsing. Meta instances are the lowest information in SA, based on which high level classes are inferred.

(3) We developed algorithms to infer general classes and abstract classes. Being able to infer classes from instances is the key to obtaining rich semantic information about SA, as classes act as the contexts in which detailed meta instances can be explained in terms of their semantics.

(4) We proposed a method to reduce the uncertainties of SA triple. This method is based on the deep analysis of the syntactic relationships between words/phrases in SA sentences.

(5) We proposed an algorithm to determine the local context of a SA sentence. The local context of a SA sentence represents the general background where most meta instances are understood. Local contexts are the basis for building knowledge needs.

The final product of SA parsing includes SA triples and local contexts of SA sentences, which are the key 'ingredients' of query construction discussed in Chapter 8.

Chapter 8
Data Warehouse Query Construction and Situation Presentation

Data warehouse (DW) queries represent the information need of a manager in the CDDP model. DW queries are constructed based on the mapping relationships between the manager's SA, the experience base, the ontology and the data warehouse. Once DW queries are constructed, situation information can be retrieved from the data warehouse through executing the queries. The retrieved situation information is presented to the manager for SA development. Part of the work presented in this chapter has been reported in one of our publications (Niu et al. 2008). Section 8.1 introduces the basic grammar of DW query languages. Section 8.2 outlines the process of query construction. The detailed processes of query construction are discussed in Sections 8.3 and 8.4. The method of presenting situation information is discussed in Sections 8.6 and 8.7.

8.1 Query Languages for Data Warehouses

Today's data warehouses can accept two types of queries: structured query language (SQL) queries, and Multidimensional Expressions (MDX) queries. SQL is the standard interactive and programming language for accessing relational data, e.g., relational tables, while MDX is for manipulation of multidimensional data, e.g., cubes. This section very briefly introduces the essentials of SQL and MDX to provide the preliminary knowledge for understanding the process of query construction.

8.1.1 Structured Query Language

There are three basic types of SQL statements: data statements for querying and modifying tables and columns, transaction statements for controlling transactions, and schema statements for maintaining database schema. The queries to be constructed in this research are used to retrieve situation information from data warehouses. Hence, we are only concerned with data statements for querying tables.

L. Niu, J. Lu, and G. Zhang: Cognition-Driven Dec. Supp. for Business Intel., SCI 238, pp. 119–139.
springerlink.com

A SQL query retrieves rows from the database tables under specified search conditions. A basic SQL query (statement) has three clauses: SELECT clause, FROM clause, and WHERE clause (Microsoft 2007c) with the following format:

SELECT column_list FROM table_name WHERE search_condition

where, *table_name* is the source table from which data will be retrieved; *column_list* is a list of columns to be selected for the result set; and *search_condition* specifies the criteria of selecting rows.

Suppose we have a relational database containing a table called EmployeeAddressTable (Hoffman 2001). This table has six columns: SSN (Social Security Number), FirstName, LastName, Address, City and State. The records of EmployeeAddressTable are list in Table 8.1.

Table 8.1 Employee Address Table

SSN	First Name	Last Name	Address	City	State
512687458	Joe	Smith	83 First Street	Howard	Ohio
758420012	Mary	Scott	842 Vine Ave.	Losant	Ohio
102254896	Sam	Jones	33 Elm St.	Paris	New York
876512563	Sarah	Ackerman	440 U.S. 110	Upton	Michigan

Some example queries based on this table are as follows.

SELECT Address, City FROM EmployeeAddressTable WHERE FirstName = 'Joe'

This query retrieves the address and City information of an employee called Joe.

*SELECT * FROM EmployeeAddressTable WHERE FirstName = 'Joe'*

This query retrieves all information of an employee called Joe.

SELECT City FROM EmployeeAddressTable

This query retrieves City information of all employees.

These SQL queries are the most basic type. There are more complex queries, for example, queries retrieving data from multiple source tables, queries with JOIN or UNION operations, and queries containing aggregate functions. This book mainly discusses the basic query type in our algorithms and systems.

8.1.2 Multidimensional Expressions

Some key concepts about MDX are as follows (Microsoft 2007b).

- Dimensions

A dimension is a perspective from which data of interest is viewed, such as products, locations and years. In data warehouses, dimensions are used to organize data with relation to different perspectives. Dimensions are constructed based on *dimension* tables. A dimension table is a relational table containing columns and rows. A *dimension* contains attributes that correspond to columns in dimension tables. The specific values of attributes in a dimension are called members. In MDX, members are defined by member expressions. A member expression contains at least two fields: the dimension name and the member name, optionally containing the attribute name. A member expression uniquely identifies a member. Some examples of member expressions are as follows.

[*Product*].[*Color*].[*Black*]

[*Product*].[*Color*].[*Silver*]

[*Location*].[*Country*].[*Australia*]

[*Location*].[*Country*].[*China*]

Where, each member expression consists of three fields: dimension name, attribute and member name.

- Measures

Measures are measurable (numeric) columns defined in fact tables, for example sales, order quantity and profit. Fact tables are also relational tables, which contain the measurements, metrics or facts of business operation. In MDX, values of measures are members of a special dimension called *Measures*. The corresponding member expressions do not contain attribute field, for example,

[*Measures*].[*Sales Amount*].

- Cubes

A cube is a set of related measures and dimensions defined by multidimensional data schema, for example, star schema and snow-flake schema. Thus, cubes are multidimensional data used to support online analytical processing (OLAP). An example of cube is shown in Figure 8.1.

This cube is defined by three dimensions: *Location, Period* and *Customer*.. Suppose the measures of this cube are *Sales Amount*, and *Order Quantity*. Then, each cell in Figure 8.1 contains a number representing the specific sales amount with relation to specific dimension members, for example, the sales amount from

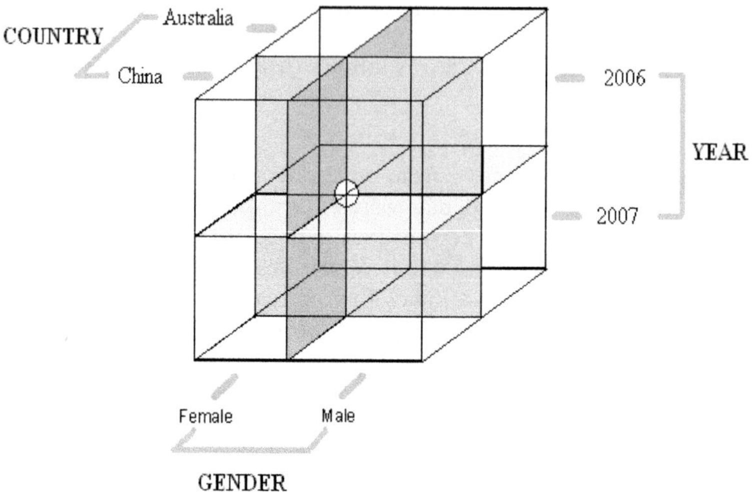

Fig. 8.1 A Cube

male customers, in 2006, in Australia. In MDX, this cell can be specified by four member expressions:

[*Customer*].[*Gender*].[*Male*]

[*Period*].[*Year*].[*2006*]

[*Location*].[*Country*].[*Australia*]

[*Measures*].[*Sales Amount*].

- Tuples

A tuple is a group of member expressions which uniquely identifies a cell. An example of a tuple is as follows.

([*Customer*].[*Gender*].[*Male*], [*Period*].[Year].[*2006*],

[*Location*].[*Country*].[*Australia*], [*Measures*].[*Sales Amount*])

- Sets

A set is an ordered set of tuples with the same dimensionality. An example of a set is as follows.

{

([*Customer*].[*Gender*].[*Male*], [*Period*].[*Year*].[*2006*],

[*Location*].[*Country*].[*Australia*], [*Measures*].[*Sales Amount*]),

([*Customer*].[*Gender*].[*Male*], [*Period*].[*Year*].[*2007*],

[*Location*].[*Country*].[*Australia*],[*Measures*].[*Sales Amount*])

}

This set contains two tuples. The difference between the two tuples is the Period dimension: one for 2006 and another for 2007. Thus it can specify two cells in the cube shown in Figure 8.1.

- Axes

In a MDX statement, multiple sets can be organized along different axes. Each query axis has a number: zero (0) for the x-axis, 1 for the y-axis, 2 for the z-axis, and so on. The following is an example of sets organized along two axes.

{([*Customer*].[*Gender*].[*Male*], [*Period*].[*Year*].[*2006*],

[*Measures*].[*Sales Amount*]),

([*Customer*].[*Gender*].[*Male*],[*Period*].[*Year*].[*2007*],

[*Measures*].[*Sales Amount*])} **ON 0**,

{([*Location*].[*Country*].[*Australia*], [*Location*].[*Country*].[*China*])} **ON 1**

We have discussed the necessary concepts of MDX queries. Now let us look at a complete example of a MDX query as follows.

SELECT

{([*Customer*].[*Gender*].[*Male*], [*Period*].[*Year*].[***2006***], [*Measures*].[*Sales Amount*]),

([*Customer*].[*Gender*].[*Male*],[*Period*].[*Year*].[***2007***],[*Measures*].[*Sales Amount*])} **ON 0**,

{([*Location*].[*Country*].[*Australia*], [*Location*].[*Country*].[*China*])} **ON 1**
FROM [*Adventure Works*]

Where, *Adventure Works* is the name of a cube. The result returned by this MDX query will be something as follows:

	Male 2006 Sales Amount	Male 2007 Sales Amount
Australia	*$448, 525*	*$949, 370*
China	*$555, 071*	*$858, 565*

8.2 Framework of Query Construction and Situation Presentation

The process of query construction and situation presentation is based on the mapping relationships between SA, experience base, ontology, data warehouse and situation, which are illustrated in Figure 8.2.

There are five layers in this figure separated by horizontal lines.

(1) Situation Awareness

In the SA layer, the manager describes his/her SA about the current decision situation and inputs into the system in the form of natural language. Each SA description is comprised of a number of SA sentences. Employing the Link Grammar Parser and relevant techniques discussed in Chapter 7, SA sentences are parsed into SA tokens. Finally, SA triples are generated by annotating the SA tokens, and local contexts of SA sentences are also determined. The SA triples and local contexts and are mapped into the ontology and the experience base.

(2) Experience Base

In this layer, the manager's experiences are represented as experience maps. An experience map consists of two sorts of concepts (cause concepts and effect concepts) and the causal relationships between them.

The local contexts of SA sentences are used to generate a knowledge need, which are then used to extract cues from the experience base. The extracted cues represent a fragment of the entire experiences, which imply potential solutions, ideas or relevant situation information. In the process of knowledge retrieval, a local context in the SA layer is directly mapped to an effect concept in the experience base layer.

(3) Ontology

Classes are defined in the ontology layer. Each class is described by a set of properties. Classes are associated with each other via two sorts of relationships: subsumption relationships and property-share relationships.

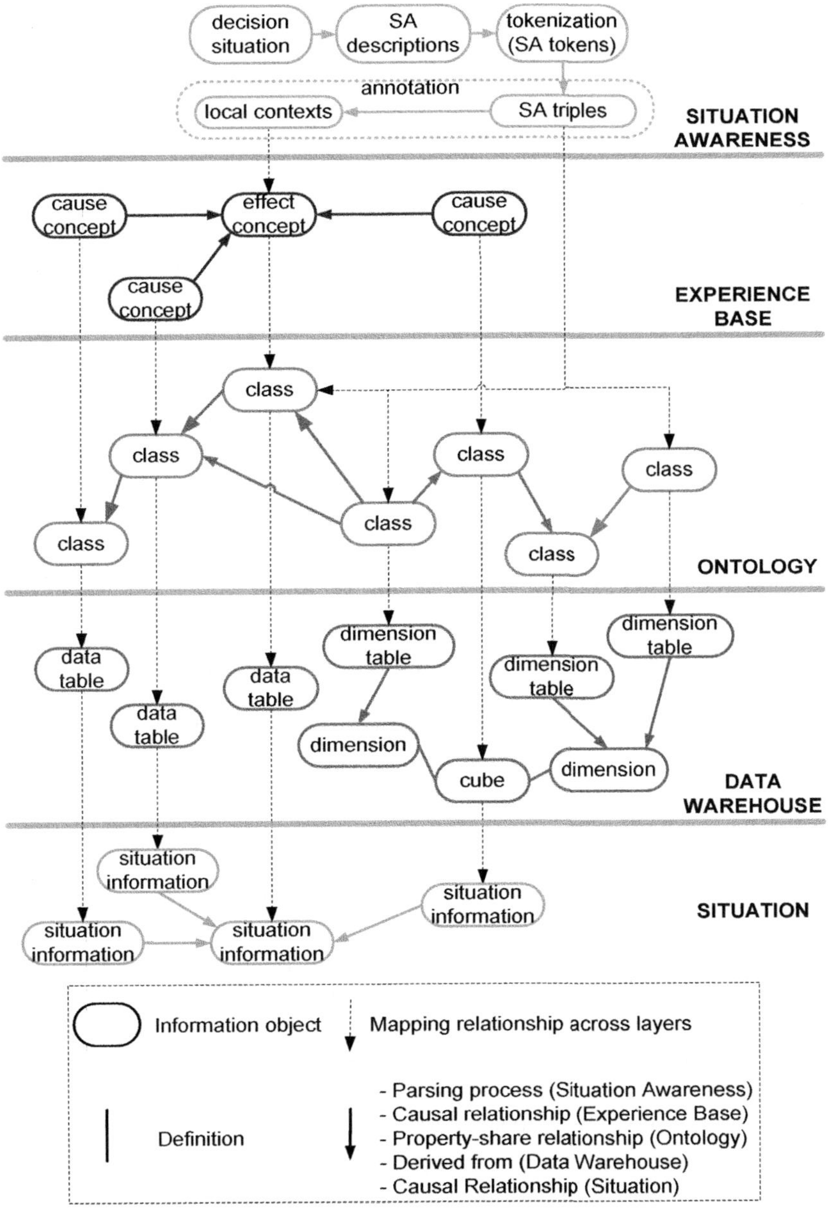

Fig. 8.2 Framework of Query Construction and Situation Presentation

A SA triple in the SA layer is mapped to a combination of a class, a property of this class and a meta instance of this property, in the ontology layer. A property-share relationship enables semantic information expressed by SA triples to be transferred from a class to another class. By this way, those classes, which do not have direct mapping relationships with SA triples, can also be indirectly mapped.

(4) Data Warehouse

In the data warehouse layer, situation information is hiding in either data tables (relational data) and/or cubes (multidimensional data). The objective of query construction is to find the situation information for situation presentation. A cube is defined by a number of dimensions and measures. Dimensions are derived from dimension tables.

A class in the ontology layer can be mapped to a data table, a cube or a dimension table in the data warehouse layer. The mapping relationship is defined during the development of the ontology.

Queries are constructed and executed against the data warehouse. During the execution of the queries, situation information is retrieved from data tables or cubes. The retrieved situation information is transferred to the situation layer for situation presentation.

(5) Situation

In this layer, the original decision situation is presented to the manager through displaying the retrieved situation information. Situation presentation is implemented in the form of different data visualization techniques, such as graphs, charts and tables. When the manager perceives and understands the presented situation information, his/her SA is developed and enriched.

8.3 Determining Query Data Sources

The process of query construction includes two basic tasks: determining query data sources and constructing query clauses. A query data source is a table (query table) or cube (query cube) from which situation information will be retrieved.

Query tables/cubes are determined according to the cues extracted from the experience base during knowledge retrieval. The extracted cues are related to the current decision situation and the manager's past experience. Cues prompt potential solutions, ideas, or information relevant to the current decision situation.

A cue consists of an effect concept and a number of cause concepts. The effect concept of a cue is considered as a subtopic of the current decision situation, while the cause concepts represent different factors affecting the subtopic. Exploring the factors of the subtopic will help the manager to gain insights into the decision situation. Thus, the data/information related to each concept in a cue should be retrieved and presented to the manager in terms decision support.

The concepts of cues are indirectly mapped to DW objects via the ontology. According the relationships between the experience base, the ontology and the data warehouse, each concept of a cue is also a class in the ontology which is further mapped to a DW object in the data warehouse. The DW object mapped with an ontology class can be of three types: a data table, dimension table or cube. Of them, only data tables and cubes are qualified query data sources.

In an application, not all ontology classes are mapped with a DW object. If a concept of a cue is mapped to a class which does not have available associated DW object, we use class similarities to search for the nearest qualified class with an associated DW object. By this way, all cue concepts can be indirectly mapped to DW objects.

For a decision situation, a number of cues might be extracted from the experience base. All these cues will be merged as a navigation knowledge. The data source for a query is determined based on the navigation knowledge. The procedure of determining the data source for a query has four steps:

Step 1. The navigation knowledge is visualized as a map and presented to the manager.

Step 2. The manager browses the navigation knowledge map and expresses interest in a specific concept.

Step 3. A class which is matched with the concept of the manager's interest is found in the class tree of the ontology.

Step 4. If the class found in Step 3 has an associated qualified DW object (data table or cube), output this DW object as the query data source. Otherwise, find the most similar qualified class through assessing class similarities and output the DW object of this new class as the query data source.

8.4 Constructing SQL Queries

We developed the algorithm ***SqlBuilder*** to construct SQL queries. *SqlBuilder* intakes a concept of the navigation knowledge and outputs the corresponding SQL query for this concept.

Algorithm 8.1. SqlBuilder

Input:

(1) A navigation knowledge concept: ***c***

(2) A set of SA triples: ***SATripleSet***

(3) An ontology

Output:

A SQL query for concept ***c***

Procedure:

See Figure 8.3.

One of the theoretical bases of *SqlBuilder* is the concept of property-share relationship. A property-share relationship between two classes enables information about one class to be transformed to another class. The information about classes here mainly refers to SA triples. The concept ***c*** in the navigation

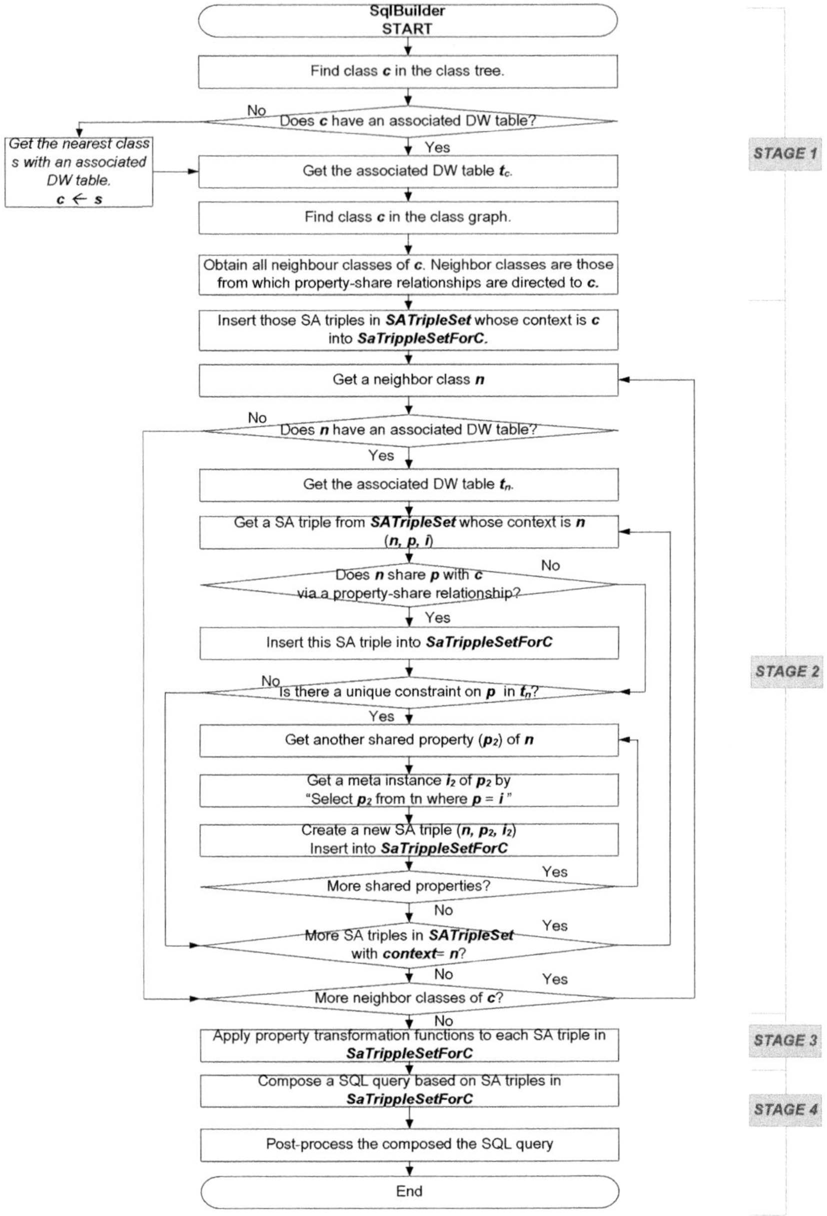

Fig. 8.3 Algorithm SQLBuilder

knowledge is also a class in the ontology. Thus, the information about ***c*** can be obtained from two sources. First, those SA triples whose contexts are ***c*** provide direct information about ***c***. Second, some SA triples whose contexts are not ***c*** can provide indirect information about ***c*** through property-share relationships.

As shown in Figure 8.3, the process of constructing a SQL query can be divided into four stages. In Stage 1, all neighbor classes of concept ***c*** are retrieved in the class graph. Concept ***c*** is first mapped to a class in the class tree and then mapped to a class in the class graph. In the class graph, a neighbor class of ***c*** is one from which a property-share relationship is directed to ***c***.

In Stage 2, shared SA triples from neighbor classes of ***c*** are determined. A shared SA triple is one whose *view* part is a shared property. Given a neighbor class ***n***, there are two ways to determine the SA triples which ***n*** shares with ***c***. First, the shared SA triples can be selected directly from the *SATripleSet*. A SA triple in *SATripleSet* is a shared one as long as its context is ***n*** and its view is defined in a property-share relationship from ***n*** to ***c***. Second, new shared SA triples can be created according to those selected from *SATripleSet*. Let ***st*** = (***n***, ***p***, ***i***) be a shared SA triple selected from *SATripleSet*, and t_n be the associated DW data table of ***n***. If ***p*** (the view of ***st***) is unique in table t_n, ***p*** will be used to retrieve a new meta instance in t_n for other shared properties of ***n***. The retrieved new meta instance is then used to created new shared SA triples for ***c***.

The SA triples that the neighbor classes share with ***c*** are transformed using the corresponding property transformation functions in Stage 3. Property transformation functions revise the wording part of each shared SA triple and replace the context with ***c***. By this means, SA triples of neighbor classes are transferred to class ***c***, which can provide indirect information about ***c***.

A SQL query is composed according to the transformed SA triples (*SaTrippleSetForC*) in Stage 4. The associated DW table t_c is used as the query table (query data source), such that a FROM clause can be constructed:

$$\textit{FROM}\ t_c$$

All columns of t_c are considered to be of interest to the current decision. Thus SELECT clause can be constructed:

SELECT *

The remaining work of the SQL query construction is to compose query criteria, i.e., WHERE clauses. The transformed SA triples in the *SaTrippleSetForC* are grouped by their view parts. Thus a number of SA triple groups are generated, in each of which SA triples have the same view. For each group of SA triples, a number of query predicates (query criteria) can be created. For example, based on SA triple the following SA triple group

{(*Mountain Bike, EnglishProductName, Mountain-100 Silver*),

(*Mountain Bike, EnglishProductName, Mountain-100 Black*)},

Two predicates are created as follows.

EnglishProductName = Mountain-100 Silver

EnglishProductName = Mountain-100 Black

The two predicates are connected using the logical operator *OR*.

EnglishProductName = Mountain-100 Silver OR EnglishProductName = Mountain-100 Black

Predicates created based on different SA triple groups are connected using the logical operator *AND*. For example, based on SA triple

(*Mountain Bike, color, red*)},

A predicate is created as follows.

Color = red

Connected with the previous two predicates using AND,

(*EnglishProductName = Mountain-100 Silver OR EnglishProductName = Mountain-100 Black*) *AND Color = red*

Thus, a WHERE clause can be composed as follows.

WHERE (*EnglishProductName = Mountain-100 Silver OR*

EnglishProductName = Mountain-100 Black) *AND Color = red*

Finally, a provisional SQL query is constructed as follows.

SELECT * FROM t_c WHERE (*EnglishProductName = Mountain-100 Silver OR EnglishProductName = Mountain-100 Black*) *AND Color = red*

In an application, the provisional query which has been composed in Stage 4 might not be fully compatible with the specific data warehouse system. Thus, Algorithm *SqlBuilder* will post-process the provisional query through tuning each clause of the query statement. The post process includes the following tasks:

- Adjust the formats of clauses. For example, data of char string type will be enclosed with quotation marks.
- Resolve conflicts between different query criteria. For example, if two *AND*-connected predicates contain columns with a unique constraint, either the provisional query will be split into two queries, or the *AND* operator will be changed to *OR*.
- Predict the number of rows which will be retrieved. The objective of constructing DW queries is to retrieve a reasonable amount of situation information for decision support. In some cases, e.g., bad-constructed queries due to poor SA descriptions, the provisional query might return much more data than the manager would expect. In these cases, the provisional query will be attached a tag which will be used to inform the manager to input more SA descriptions. The provisional query could also return empty data set. In this case, this query will be

removed. In order to predict the number of rows, the provisional query will be submitted to the data warehouse for execution.

- Remove duplicate queries.

After post-process, *SqlBuilder* will output the following query.

SELECT * FROM t_c WHERE (*EnglishProductName =' Mountain-100 Silver' OR EnglishProductName = 'Mountain-100 Black') AND Color = 'red'*

Note that *Mountain-100 Silver* and *Mountain-100 Black* are enclosed with quotation marks according to their data type (*nvarchar*) in the definition of t_c.

In the current version, *SqlBuilder* can generate the basic type of query. We are currently working to enhance the function of *SqlBuilder*. The new version will be able to construct more complex queries, e.g., supporting *JOIN* of multiple tables.

8.5 Constructing MDX Queries

We developed the algorithm ***MdxBuilder*** to construct MDX queries. *MdxBuilder* intakes a concept of a navigation knowledge and outputs the corresponding MDX query.

Algorithm 8.2. MdxBuilder

Input:

(1) A navigation knowledge concept: ***c***

(2) A set of SA triples: ***SATripleSet***

(3) An ontology

Output:

A MDX query for concept ***c***

Procedure:

See Figure 8.4.

As shown in Figure 8.4, the process of constructing a MDX query can be divided into four stages.

- Stage 1

Given a navigation knowledge concept ***c***, the corresponding DW cube ***m*** is determined and all dimensions of ***m*** are retrieved.

- Stage 2

A sub-algorithm *CreateMemberExpression* is called to create member expressions for every dimension of m.

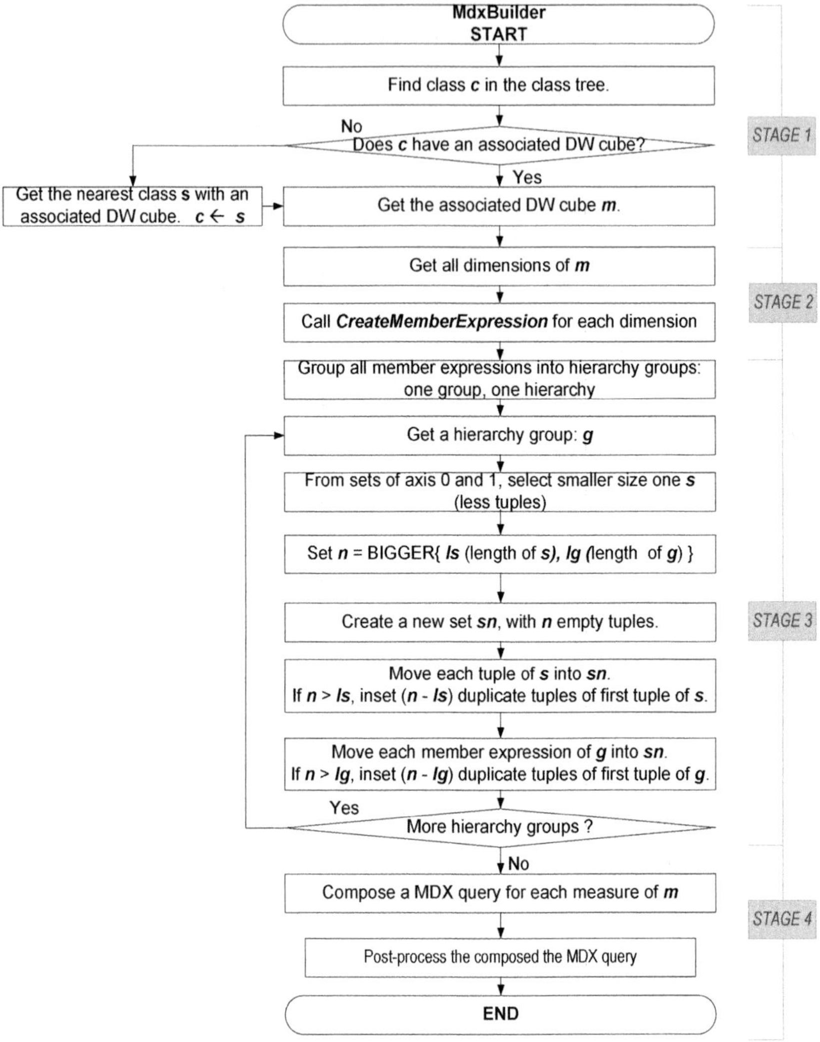

Fig. 8.4 Algorithm MDXBuilder

Algorithm 8.3. CreateMemberExpression

Input:

(1) A dimension ***d***

(2) A set of SA triples: ***SATripleSet***

Output:

A set of member expressions of ***d***

Procedure:

See Figure 8.5.

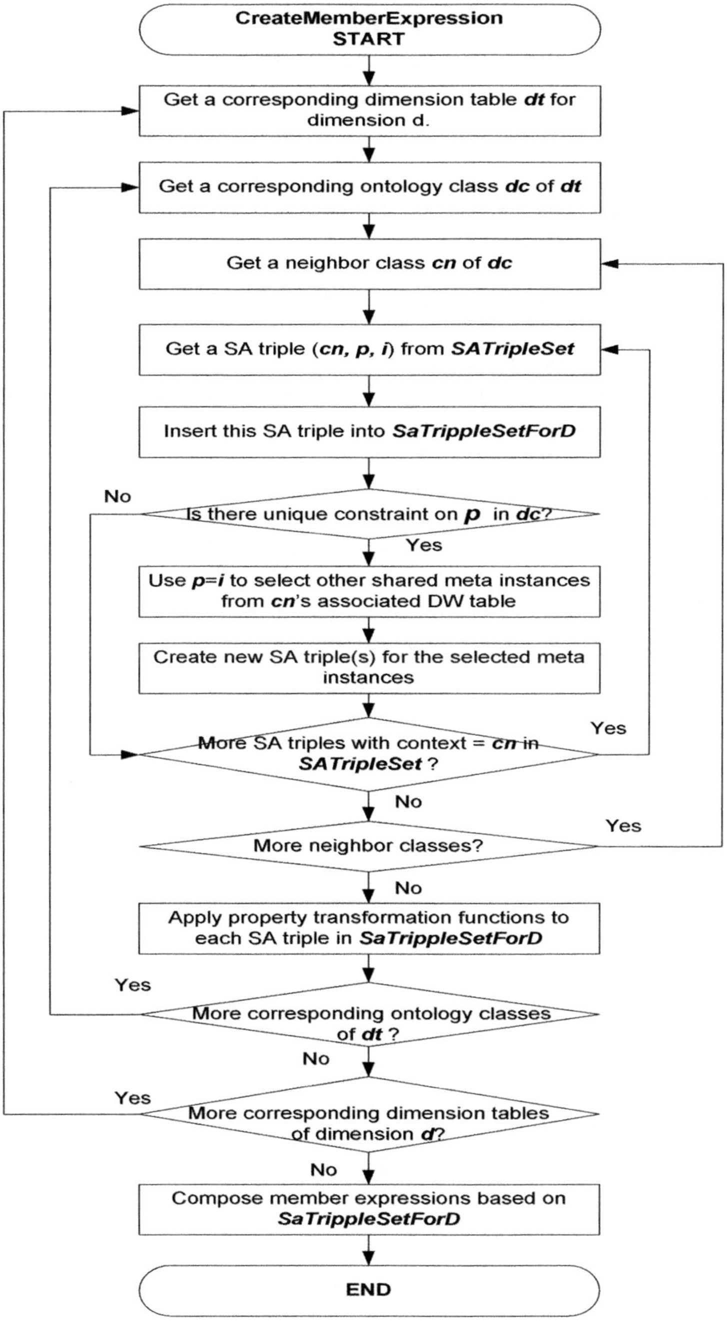

Fig. 8.5 Algorithm CreateMemberExpression

The algorithm *CreateMemberExpression* is also based on the concept of the property-share relationship. In the data warehouse, dimensions are related to dimension tables. Dimension tables are mapped to classes in the ontology according to the mapping relationships between the ontology and the data warehouse (Figure 8.2).

A dimension table ***dt***, corresponds one or more classes in the ontology. Let ***dc*** be a corresponding ontology class of ***dt***. Just as algorithm *SqlBuilder*, *CreateMemberExpression* retrieves the shared SA triples in *SATripleSet* and creates new shared SA triples based on retrieved meta instances for class ***dc***. *CreateMemberExpression* collects all shared SA triples into set ***SaTrippleSetForD*** for all corresponding classes of ***dt*** and for all corresponding dimensional tables of ***d***.

The member expressions of dimension ***d*** are composed based on SA triples in ***SaTrippleSetForD***. The relationships between the dimension attributes and the columns of dimension tables can be extracted from the schema of the data warehouse. Thus, properties of classes are mapped to attributes of dimensions according to the mapping relationships between ontology classes and the DW dimension tables. The mapping relationships between dimension attributes and class properties enable each SA triple in ***SaTrippleSetForD*** to be transformed into a member expression of dimension ***d***. Let us look at an example.

Example 8.1. Constructing Member Expression

SA Triples:

(*Mountain Bike, EnglishProductName, Mountain-100 Silver*)

(*Road Bike, EnglishProductName, Road--30 Black*)

(*Mountain Bike, Color, Blue*)

(*Customer, Location, France*)

(*Customer, Location, Canada*)

Member Expressions which are composed based on the above SA triples:

[*Product*].[*Model Name*].[*Mountain-100 Silver*]

[*Product*].[*Model Name*].[*Road--30 Black*]

[*Product*].[*Product Color*].[*Blue*]

[*Customer*].[*Country*].[*France*]

[*Customer*].[*Country*].[*Canada*]

- Stage 3

MDX tuples and sets are generated in this stage. Each MDX set is comprised of a number of tuples with the same dimensionality. Tuples are constructed using member expressions created in Stage 2 and then evenly distributed into the two sets.

Based on the member expressions generated in Example 8.1, two MDX sets can be constructed as follows.

Axis 0 set:

{([*Product*].[*Model Name*].[*Mountain-100 Silver*], [*Product*].[*Product Color*].[*Blue*]), ([*Product*].[*Model Name*].[*Road--30 Black*], [*Product*].[*Product Color*].[*Blue*]) }

Axis 1 set:

{([*Customer*].[*Country*].[*France*]), ([*Customer*].[*Country*].[*Canada*])}

Based on the same collection of tuples, different MDX sets can be constructed which are equally reasonable in terms of MDX syntax. For example, based on the member expressions generated in Example 8.1, the following MDX sets can also be constructed.

Axis 0 set:

{([*Product*].[*Model Name*].[*Mountain-100 Silver*]),

([*Product*].[*Model Name*].[*Road--30 Black*]) }

Axis 1 set:

{([*Customer*].[*Country*].[*France*], [*Product*].[*Product Color*].[*Blue*]),

([*Customer*].[*Country*].[*Canada*], [*Product*].[*Product Color*].[*Blue*])}

Based on a collection of member expressions, there are many ways to generate the corresponding MDX sets. Different MDX sets will results in different MDX queries which will return different information. There are no formal rules which can be used to determine the number of MDX sets and the dimensionality of tuples. Algorithm *MdxBuilder* will initially generates two sets for a MDX query: axis 0 set (columns) and axes 1 set (rows), and randomly generate a dimensionality of tuples. However, in the prototype FACETS, the manager are allowed to use up to three axes and customize the dimensionality according to his/her requirements.

- Stage 4

This is the final stage where a complete MDX query is constructed by assembling tuples, the query cube and measure. MdxBuilder constructs a MDX query for every measure of a cube. Let ***r*** be a measure of cube ***m***. The following is an example of a finished MDX query.

SELECT

{([*Product*].[*Model Name*].[*Mountain-100 Silver*], [*Product*].[*Product Color*].[*Blue*]), ([*Product*].[*Model Name*].[*Road--30 Black*], [*Product*].[*Product Color*].[*Blue*]) } ON 0,

{([*Customer*].[*Country*].[*France*]), ([*Customer*].[*Country*].[*Canada*])} ON 1

FROM [*m*]

WHERE [*r*]

Note that in the above MDX query, measure ***r*** is used as a WHERE clause, instead of the member of the special dimension *Measure*.

8.6 Navigation-Knowledge-Guided Situation Presentation

We use navigation knowledge to guide the process of situation presentation. This method is depicted in Figure 8.6. The relationships between situation information and environment data are established by DW queries. When DW queries are executed against the DW, corresponding situation information can be retrieved. A piece of situation information might correspond to different data in the DW. In Figure 8.6, solid directed lines are used to denote these direct relationships.

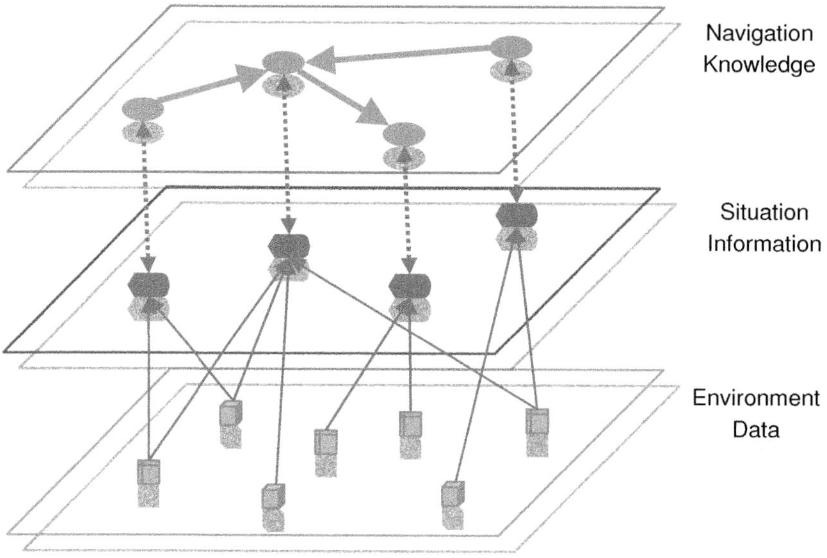

Fig. 8.6 Navigation-knowledge-guided Situation Presentation

The dashed directed lines denote indirect relationships between situation information and navigation knowledge. Navigation knowledge is visualized as maps on a graphical user interface (GUI). The representation of navigation knowledge also consists of concepts and causal relationships. Each concept implies a kind of information which reflects an aspect of the decision situation. If the manager is interested in a specific concept in the navigation knowledge, he/she can request the system to show related situation information. For example, the request can be made by mouse-clicking onto a concept on the navigation knowledge map. When the system receives an information request, the corresponding queries will be triggered and executed against the data warehouse. Relevant data is then retrieved from the data warehouse (environment data). The retrieved data is processed using data analysis techniques and then presented to the manager on the GUI.

The general pattern of human-machine interaction in the navigation-knowledge guided situation presentation is as follows.

A navigation knowledge map is presented →

(The manager) Browse the experience →

Find a concept interesting →

Click this concept for associated information →

Detailed information is presented →

Perceive the detailed information →

Return back the navigation knowledge map →

Browse the experience →

Find another concept interesting →

...

The navigation knowledge is the combination of all cues relevant the current decision situation. Cues prompt possible ideas, clues or solutions to current decision situation. A cue is retrieved according to a concept (an element of the knowledge needs related to the current decision situation) from the experience base. A decision situation might have many related concepts. A cue is generated only based on one concept. Hence, a cue can only reflect part of the decision situation. The navigation knowledge is the integration of all related cues of a decision situation. Therefore, the visualization of navigation knowledge presents a complete *big picture* of the current decision situation to the manager.

The navigation-knowledge-guided situation presentation also allows managers to add extra concepts and causal relationships to the generated navigation knowledge. For example, suppose the navigation knowledge shows that

PRODUCT and ADVERTISEMENT are the key factors to SALES. If a manager thinks GOVERNMENT POLICY is also playing a role, he/she can supply this new concept and the relationship with SALES to the system. The navigation knowledge map and the corresponding DW queries will be updated accordingly. The navigation-knowledge-guided situation presentation is implemented in FACETS (Chapter 7).

8.7 Data Analysis and Situation Presentation

The retrieved data is not always easily understood by managers. For instance, a DW query might return many sales data across different locations, products and time periods. Thus, it is necessary to perform data analysis on the retrieved data before presenting to the manager.

The data analysis techniques discussed in this section are not basic components of the CDDP model. However, in terms of application, they are a substantial part of the integral system built based on the CDDP model. Thus, we briefly discuss each of them as follows.

There are three basic categories of data analysis techniques in today's BI applications.

(1) Relational Data Reporting

Relational data reporting (RDR) is the most well-developed and widely-used data analysis technique in the BI area. RDR is based on executing SQL queries against OLTP systems. The data being reported by RDR has a very simple structure: relational tables. The visual forms which are used to present data are very varied, such as list tables, cross tables, bar graphs, line graphs, pie charts, pie slices and divided bars. RDR is one of the most basic analysis functions of most commercial BI tools such as BO, Cognos, Hyperion, Microstrategy, SAS, SAP, Microsoft and ORACLE.

(2) Multidimensional Data Reporting

Multidimensional data reporting (MDR) was born with the development of OLAP systems. The structure of multidimensional data (cubes) can be star schema, snowflake schema or hybrid schema, which are more complex than relational tables. MDR tools usually support ad hoc data analysis whereby users can undertake customized analytics such as dicing, slicing, drilling down, drilling up and drilling through.

(3) Intelligent Data Analysis

Intelligence data analysis (Chidambaram 1996) refers to data analysis methods which involve artificial intelligence techniques. Examples of these techniques are data mining, data fusion, evolutionary algorithms, machine learning, neural nets, fuzzy logic and pattern recognition. In practical BI applications, data mining is growing as one of the most significant IDA techniques. Data mining is integrated with many commercial BI products, such as Intelligent Miner in DB2, ODM in Oracle, Enterprise Miner in SAS and Teradata Warehouse Miner in Teradata.

8.8 Summary

This chapter presents the methods and algorithms used for query construction and situation presentation. The main points of the proposed methods and algorithms are as follows.

(1) A framework for query construction and situation presentation was proposed. This framework establishes the mapping relationships between SA, experience base, ontology, data warehouse and situation.
(2) Based on the proposed framework, the algorithm *SqlBuilder* was developed to construct SQL queries.
(3) Based on the proposed framework, the algorithm *MdxBuilder* was developed to construct MDX queries.
(4) A method of situation presentation guided by navigation knowledge was discussed.

This chapter together with Chapters 6 and 7 are the technical part of this research. The technical part consists of different algorithms and methods which are developed according the CDDP model proposed in the theoretical part (Chapter 5). In the next chapter, we will report on the prototype system developed to evaluate the proposed techniques.

Part IV
Systems and Applications

Chapter 9
A Cognition-Driven Decision Support System: FACETS

Using the techniques discussed in Chapters 6, 7 and 8, we developed a prototype system, FACETS, as an implementation of the CDDP model. It allows a manager to describe his/her SA in the form of English. Based on the domain knowledge, FACETS parses the manager's SA and constructs data warehouse queries. Queries are submitted to the data warehouse for retrieving relevant situation information. The retrieved situation information is presented to managers according to the navigation knowledge extracted from the manager's experience. The goal of FACETS is to assist managers to develop and enrich their SA for decision making.

9.1 The Development Environment

We developed FACETS on a desktop computer with an AMD Athlon 64 2GHz Processor and 1 G RAM. The software environment is as follows.

- Operating System: Microsoft Windows XP Professional Version 2002 with Service Pack 2
- Programming Tool: Microsoft Visual Studio 2005 Professional Edition
- Programming Language: C/C++
- Database Server: SQL Server 2005 Enterprise Edition
- OLAP Server: Microsoft Analysis Service 2005

9.2 The Architecture of FACETS

The prototype FACETS is comprised of eight subsystems: *Data Warehouse System, Ontology Management, Experience Management, Situation Awareness Management, Situation Awareness Parsing, Situation Awareness Annotating, Query Builder* and *Situation Presentation.* Each subsystem includes a number of components which are shown in Figure 9.1.

L. Niu, J. Lu, and G. Zhang: Cognition-Driven Dec. Supp. for Business Intel., SCI 238, pp. 143 – 156.
springerlink.com

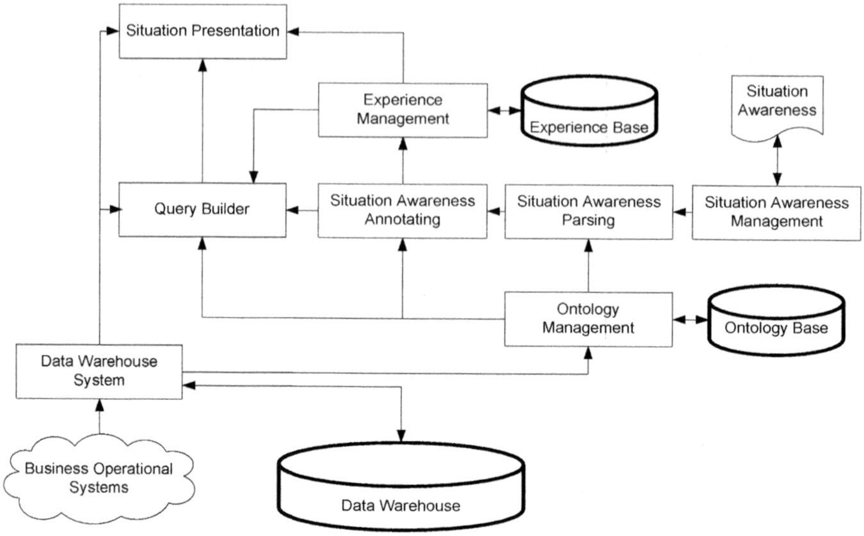

Fig. 9.1 The Architecture of FACETS

FACETS was built on the platform of a data warehouse, which includes a central data warehouse and other components for extracting data from business operational systems, managing meta data and analyzing data.

The functions of FACETS are based on an ontology and manager's experiences. Users (business managers or IT engineers) use the *Ontology Management* subsystem to create and edit the ontology. The data of the ontology is stored in the *Ontology Base*. The *Ontology Management* subsystem also provides a set of functions for operating the ontology, such as traversal, editing and computing class similarities. The *Experience Management* subsystem provides functions for creating, storing, editing managers' experience and extracting navigation knowledge for the *Situation Presentation* subsystem.

The *Situation Awareness Management* subsystem is the interface via which managers input their SA. SA is stored in form of persistent hard disk files and then sent to the *Situation Awareness Parsing* subsystem for extracting SA tokens. SA tokens are the inputs of the *Situation Awareness Annotating* subsystem responsible for generating SA triples. SA triples are the basis of query construction which is done by the *Query Builder* subsystem. The processes of parsing, annotating SA, and constructing queries are based on the domain ontology. Thus, *Situation Awareness Parsing*, *Situation Awareness Annotating* and *Query Builder* have inputs from the *Ontology Management* subsystem. In addition, constructing queries also relies on experience which comes from the *Experience Management* subsystem. The constructed queries are executed in the *Situation Presentation* subsystem for the retrieval of situation information. The retrieved situation information is finally presented to managers.

9.3 Subsystems of FACETS

9.3.1 Data Warehouse System

The role of the *Data Warehouse* subsystem is three-fold. First, it stores all business data which is the data source of situation information retrieval. Second, the business data is used for creating the ontology. Last, FACETS uses the DW subsystem to store some configuration information.

The major tasks of developing the DW subsystem include designing the data warehouse model and populating DW tables, dimensions and cubes. FACETS is not domain-specific and can be configured to be used in any decision-making domain. Thus, in this chapter, we will not report on the technical details of building up a data warehouse. A specific data warehouse (about a manufacture company) was implemented in the experiment part of this research.

9.3.2 Ontology Management

The *Ontology Management* (OM) subsystem provides a graphical interface for users (mainly IT engineers) to create ontology structures, edit classes and relationships. It also opens an application programming interface (API) for other subsystems (*Situation Awareness* Parsing, *Situation Awareness Annotating* and *Query Builder*) to access ontology manipulation functions. OM stores ontology data in an *Ontology Base* which is based on comma-separated-values files. Figure 9.2 is a screenshot of the OM subsystem.

Using the OM subsystem, users can build ontologies from scratch. That means all terms which represent classes as well as the definitions of classes are coded manually. This method is appropriate for small scale ontologies, for example, those with twenty or less classes.

It would be very time-consuming to define every class for large ontologies. The OM subsystem therefore offers an *import* utility to assist users to create ontologies efficiently. Using this utility, users can firstly import relevant meta data (tables and their definitions) from a data warehouse. The imported meta data are the initial class definitions which act as the roughcast for fine class defining. Users can explore the initial class definitions and make appropriate revisions. Based on our experience, the revisions/changes are mainly made on class names. Most table definitions can be retained as class definitions. This method speeds up the engineering process of ontology development. There are also widely used tools for ontology development, for example Protégé [7], which support standard ontology language, such as RDF and OWL. However, in the current version of FACETS, it does not support ontologies created based on other tools.

The OM subsystem also provides functions for *Situation Awareness Parsing, Situation Awareness Annotating* and *Query Builder.* Examples of these functions are as follows.

- Check if a term is a valid class label
- Obtain the parent class for a given class

---

[7] More detailed website of Protégé is at http://protege.stanford.edu/

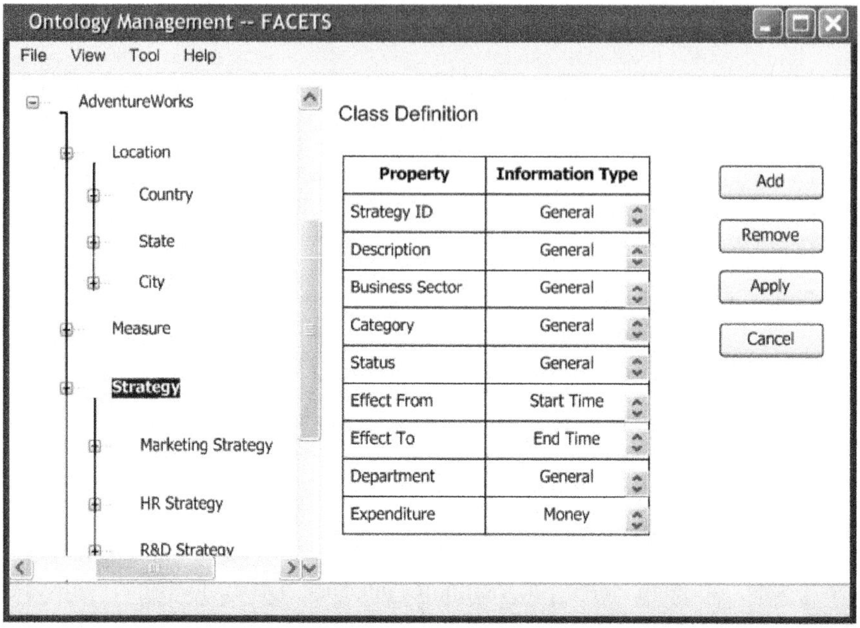

Fig. 9.2 Ontology Management

- Obtain all child classes for a given classes
- Compute the similarities between the two classes
- Obtain the associated DW object for a given class
- Determine the depth of the class tree
- Obtain the neighbor classes for a given class
- Compute the distance to the nearest leaf node for a given class

Some of these functions are implemented based on a C++ tree class called tree.hh[8].

9.3.3 Experience Management

The *Experience Management* (EM) subsystem has a graphical interface for users (mainly IT engineers) to create the structure of experience, edit experiences and relationships. It also opens an API for subsystems *Query Builder* and *Situation Presentation* to access relevant experience manipulation functions. EM stores the data of experiences in the *Experience Base* which is based on comma-separated-values files.

Figure 9.3 is a screenshot of the EM subsystem.EM is not a tool intended for eliciting fresh experience (mental models) from managers. Experience elicitation

[8] Tree.hh is available at http://www.aei.mpg.de/~peekas/tree/

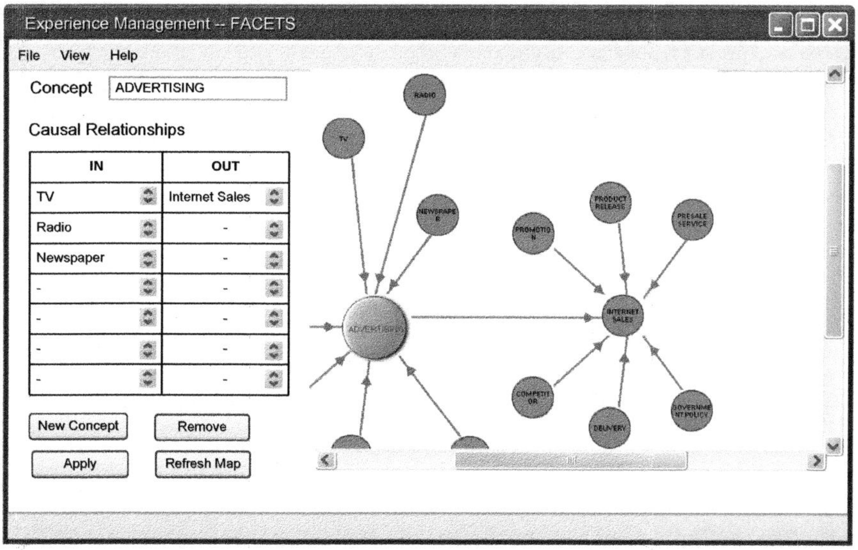

Fig. 9.3 Experience Management

is a survey process during which managers are interviewed by answering questions or describing concepts. Experience elicitation is essentially a manual process. Therefore, to build the experience base for FACETS, the experience, as a kind of knowledge, must have been already elicited and recorded. The function of EM is intended for the transformation of the form of experience, for example, from manuscripts to computer files, which then becomes understandable for FACETS.

Experiences can be visualized as experience maps in EM as shown in Figure 9.3. Users can click onto a concept and select the causal relationships related to this concept. Users can then edit this concept and relationships. Typical operations for creating experience are as follows.

- Create new concepts
- Change names of concepts
- Create new causal relationships
- Delete concepts and causal relationships
- Visualize experience

The EM subsystem also implements APIs for other subsystems. The algorithm *CueExtraction* is implemented in EM, which extracts cues for the *Query Builder* subsystem on the basis of the output of the *Situation Awareness Annotating* subsystem. The algorithm *Navigator* is implemented in EM, which generates navigation knowledge for the *Situation Presentation* subsystem on the basis of the extracted cues.

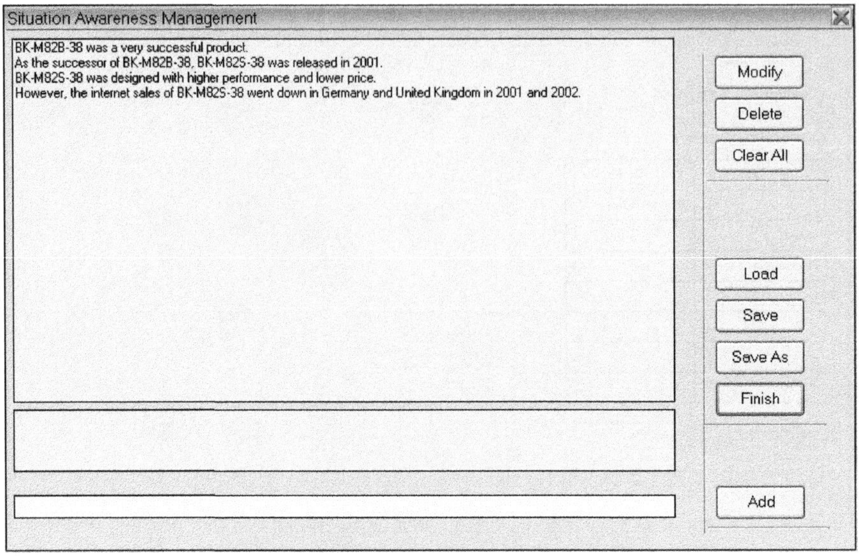

Fig. 9.4 Situation Awareness Management

9.3.4 Situation Awareness Management

The *Situation Awareness Management* (SAM) subsystem is one of the two major interfaces (another is the *Situation Presentation* subsystem) open to managers for interaction during decision making, e.g., describing SA and receiving situation information (Figure 9.4). Interfaces of *Data Warehouse, Ontology Management* and *Experience Management* are mainly for IT engineers to maintain FACETS.

Managers can input their SA via SAM. SAM stores the data of SA in the computer memory; however managers can save SA as a persistent file on the hard disk and load it for reuse. The APIs provided by SAM are open to the *Situation Awareness Parsing* subsystem, via which managers' SA descriptions are transferred for parsing using the Link Grammar Parser.

As shown in the screen shot of SAM (Figure 9.4), a number of editing functions are provided for managers to describe their SA, as follows.

- Add a new SA sentence
- Modify a SA sentence
- Delete a SA sentence
- Empty current SA
- Save all SA sentences onto hard disk file
- Load a saved SA file

Generally speaking, a complete decision process is comprised of a number of decision cycles. At the beginning of each decision cycle, managers will describe and input their current SA. Based on managers' SA, FACETS seeks and retrieves relevant knowledge and situation information, and then presents situation information back to managers. At the end of a decision cycle, managers' SA is developed or enriched when situation information is perceived and understood. During the iteration of decision cycles, SAM accumulates managers' SA by keeping historical SA inputs. However, if circumstances change, managers are allowed to modify, replace or discard the past SA.

9.3.5 Situation Awareness Parsing

The *Situation Awareness Parsing* (SAP) subsystem intakes SA sentences from *Situation Awareness Management* and parses them into SA tokens. Algorithms *NumericPlainParser*, *LiteralPlainParser*, *SemanticParser* and *LocalContextDetermination* are implemented in SAP. SAP is mainly for internal call by the *Situation Awareness Annotating* subsystem. However, we also developed a GUI in SAP for managers to monitor and control the parsing process (Figure 9.5).

On the GUI of SAP, all new SA sentences waiting to be parsed are listed. The SA sentences being parsed and the overall progress of the parsing task are dynamically updated. The SA tokens are output as parsing results. Managers can start or abort a parsing task in progress; they can also re-do the whole parsing task. In Figure 9.5, a SA sentence is being parsed: "*BK-M82S-38 was designed with higher performance and lower price.*"

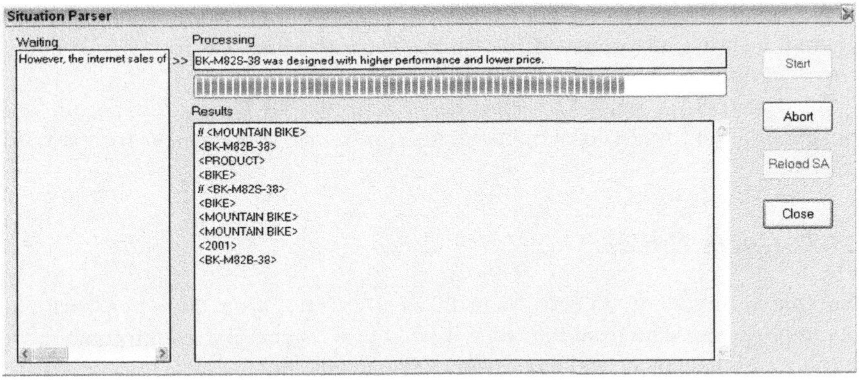

Fig. 9.5 Situation Awareness Parsing

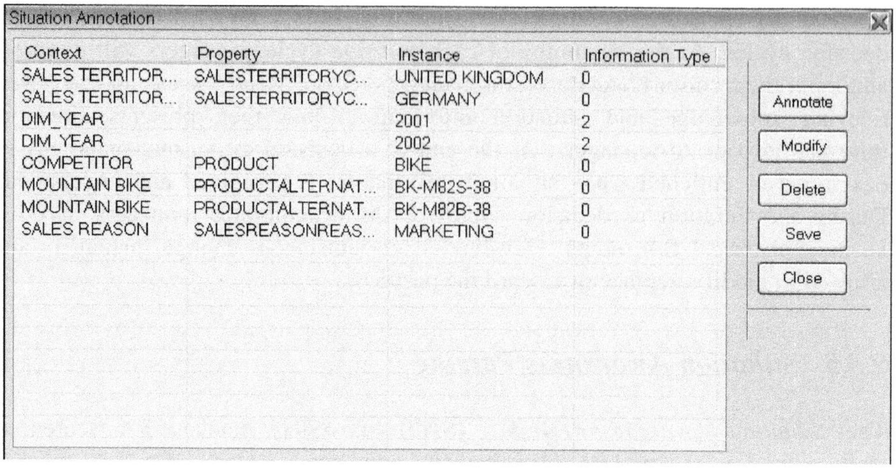

Fig. 9.6 Situation Awareness Annotation

9.3.6 *Situation Awareness Annotating*

The output of *Situation Awareness Parsing* is input into the *Situation Awareness Annotating* (SAA) subsystem. In SAA, SA tokens are further processed. SAA is mainly for internal call by subsystems *Query Builder* and *Experience Management*, but it also has a GUI for managers to amend SA annotation when necessary (Figure 9.6).

SAA annotates SA tokens and then generates SA triples. The uncertainties of SA triples are also reduced by SAA. However, if the uncertainty of a SA triple cannot be eliminated completely, managers have opportunities to participate in the process of annotating via the SAA GUI. For example, managers can delete inappropriate SA triples or modify the context, view and wording of a SA triple. Figure 9.6 shows all generated SA triples, each of which has four fields: context, property (view), instance (wording) and information type.

Subsystem SAA sends the generated SA triples to *Query Builder* for query construction, and sends local contexts to *Experience Management* for knowledge retrieval.

9.3.7 *Query Builder*

The *Query Builder* (QB) subsystem takes SA triples from *Situation Awareness Annotating*, meta data from the *Data Warehouse* system, classes information from Ontology Management and cues from Experiment Management. Based on these inputs, data warehouse queries are constructed. The constructed queries are then fed to the *Situation Presentation* subsystem for retrieving situation information.

Query building is a completely automatic process, but for the purpose of illustrating research, QB also has a GUI for users (Figure 9.7). On this GUI, users can control the process of query construction. They can also edit the generated queries.

Queries are attached to specific cue concepts. QB has a list of all concepts. For each concept, the attached queries are also shown in the GUI. For example, in Figure 9.7, the concept MOUNTAIN BIKE is selected from the concept list. There is only one query (shown in left box) related to this concept. The detail of this query is shown in the right box.

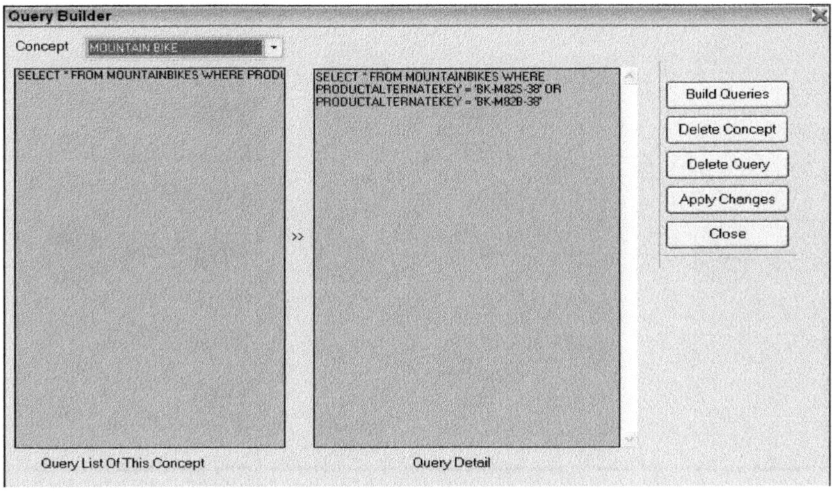

Fig. 9.7 Query Builder

9.3.8 Situation Presentation

The *Situation Presentation* (SP) subsystem is another GUI intended for managers' use in addition to *Situation Awareness Management*. SP presents situation information to managers for situation assessment. Behind the GUI is the implementation of two reporting technologies: SQL reporting and MDX reporting.

In the SP subsystem, managers will be firstly presented with a navigation knowledge map regarding their experience and the current SA description. An example of a navigation knowledge map is shown in Figure 9.8. In the navigation knowledge map, circles denote concepts and directed lines denote causal relationships. Managers can browse the whole map, which can help them recall past experience and correlate past experience with the current decision situation. The examination of navigation knowledge maps stimulates managers to identify possible ideas, factors or solutions for the current decision situation.

If managers are interested in a concept in the navigation knowledge map, they can have instant access to the related data behind this concept by simply clicking on the concept (visualized as a circular button). When a concept is clicked, SP will evoke the appropriate reporting module (SQL or MDX) to retrieve situation information relevant to this concept and generate a report for presentation. If the concept being clicked corresponds to a relational table, the SQL reporting module

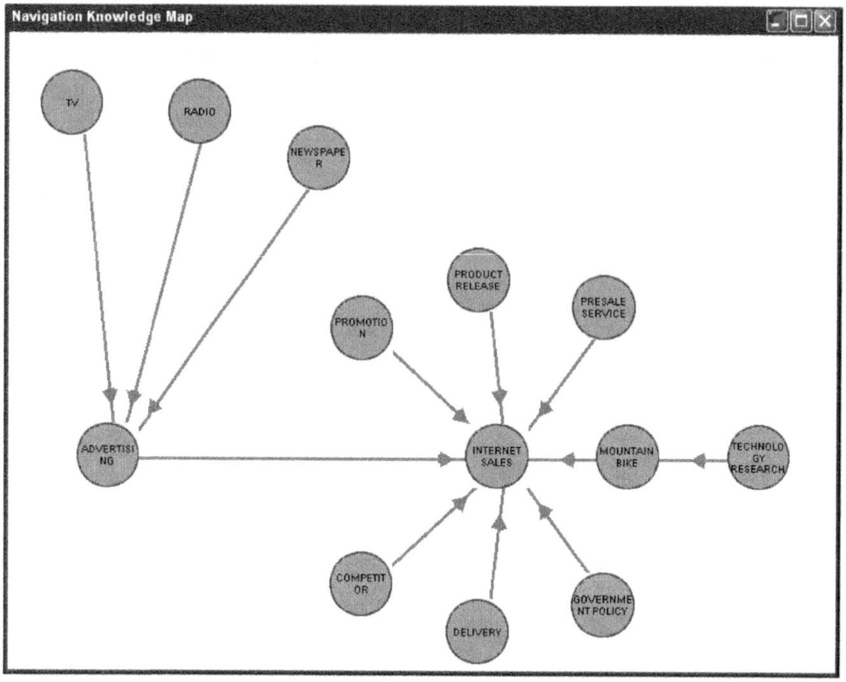

Fig. 9.8 Navigation Knowledge Map

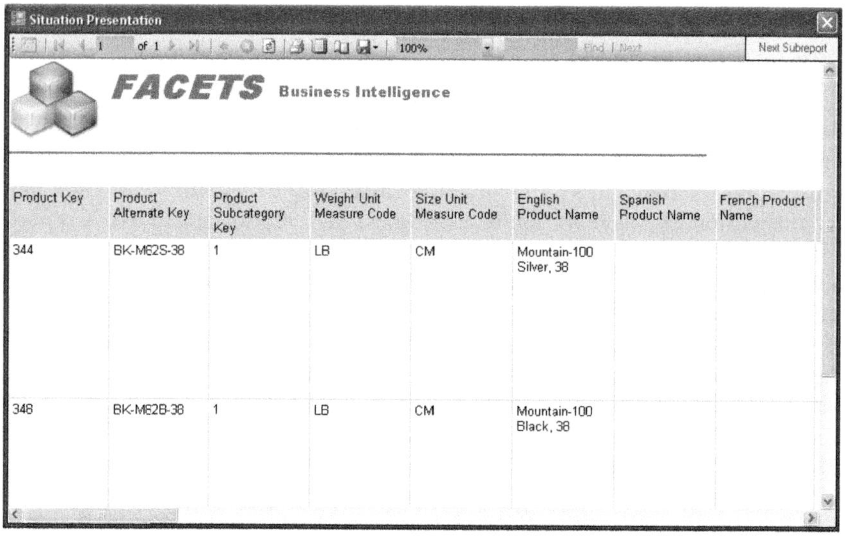

Fig. 9.9 SQL Reporting

will be called. Accordingly, a SQL report is generated and presented. Figure 9.9 shows a SQL report generated by the SQL reporting module for the concept MOUNTAIN BIKE.

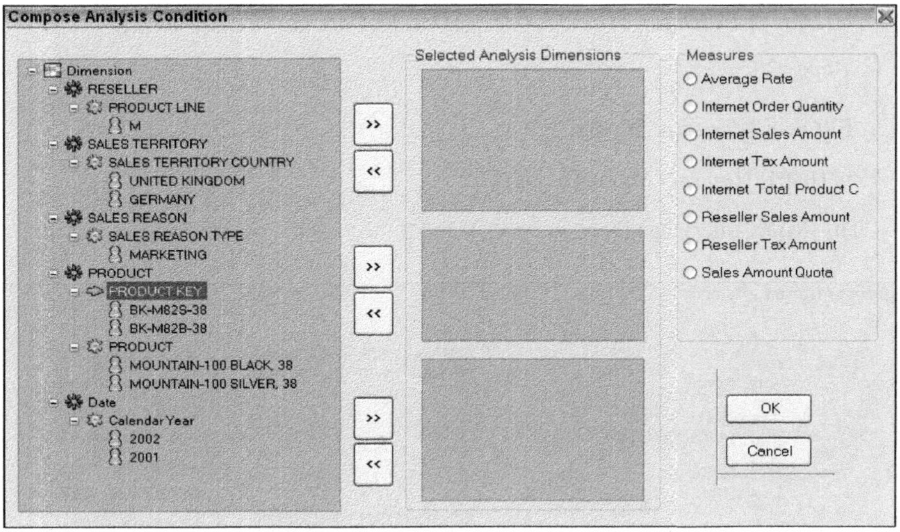

Fig. 9.10 MDX Analysis Condition Customization

If the concept being clicked corresponds to a cube, the MDX reporting module will be called. The MDX reporting module will firstly require managers to choose the analysis condition. Analysis conditions consist of dimensions, attributes, members and measures. As shown in Figure 9.10, a number of relevant dimensions such as RESELLER, SALES TERRITORY, PRODUCT and DATE are available for selection. Under each dimension are a number of attributes; under each attribute are a number of members. For example, Sales Territory Country is an attribute for dimension SALES TERRITORY; United Kingdom and Germany are members for attribute Sales Territory Country.

Managers can organize the analysis conditions of their selection along three axes (X, Y and Z) displayed as the top box, middle box and bottom box under label *Selected Analysis Dimensions* in Figure 9.10. The available measures defined with the corresponding cube are also shown in the GUI, such as Average Rate, Internet Order Quantity and Internet Sales Amount.

A finished analysis condition customization is shown in Figure 9.11, which results in a MDX query as follows.

SELECT

{([*PRODUCT*].[*PRODUCT*].[*Mountain-100 Silver, 38*]),

([*PRODUCT*].[*PRODUCT*].[*Mountain-100 Black, 38*])

} ON 0,

{([*SALES TERRITORY*].[*SALES TERRITORY COUNTRY*].[*United Kingdom*]),

([*SALES TERRITORY*].[*SALES TERRITORY COUNTRY*].[Germany])

} ON 1,

{([*Date*].[*Calendar Year*].[*2001*]), ([*Date*].[*Calendar Year*].[*2002*])

} ON 2

FROM [*Adventure Works*]

WHERE [*Internet Sales Amount*]

This MDX query has there axes: 0, 1 and 2.

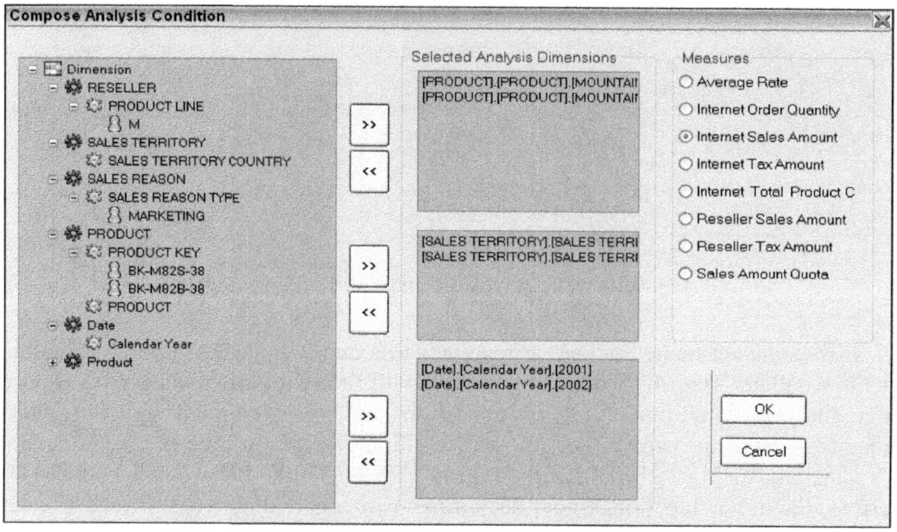

Fig. 9.11 Finished MDX Analysis Condition Customization

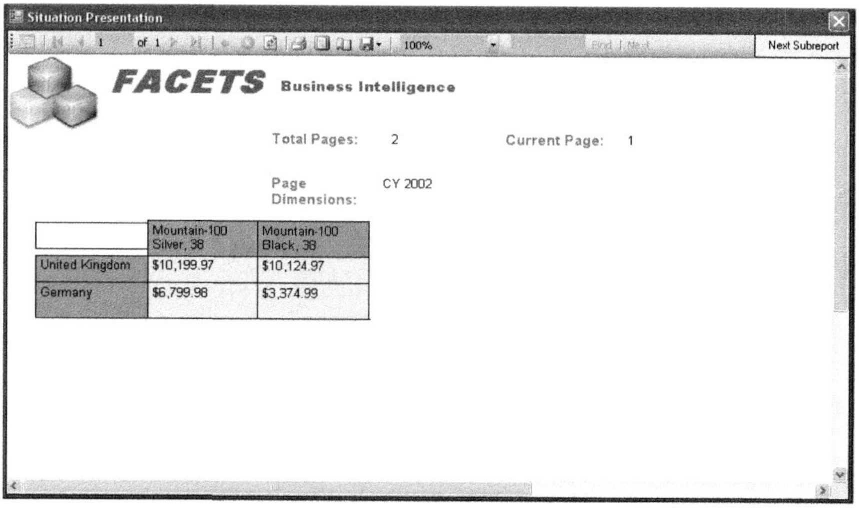

Fig. 9.12 A MDX Report

When the manager finishes customizing the analysis conditions, the corresponding MDX query will be generated and submitted to the data warehouse to retrieve relevant situation information. The retrieved situation information is presented to the manager in the form of MDX report. Figure 9.12 shows a MDX report generated by the MDX reporting module according to the analysis condition defined in Figure 9.11. This MDX report shows the internet sales amount across the United Kingdom and Germany for different product models in the year 2002. There are two pages of reports generated for 2002 and 2001 respectively. Managers can click on the *Next Report* button to see different pages.

9.4 The Cognition-Driven Decision Process Based on FACETS

FACETS has eight subsystems in total. However, in terms of decision support, only two of them are interfaces for managers to interact with the system for decision making: *Situation Awareness Management* and *Situation Presentation* (Figure 9.13). Other subsystems are mainly for IT engineers to manage and configure the system. In real applications, the whole process of SA parsing, annotating, constructing queries and retrieving situation information can be fully automatized without managers' manual intervention.

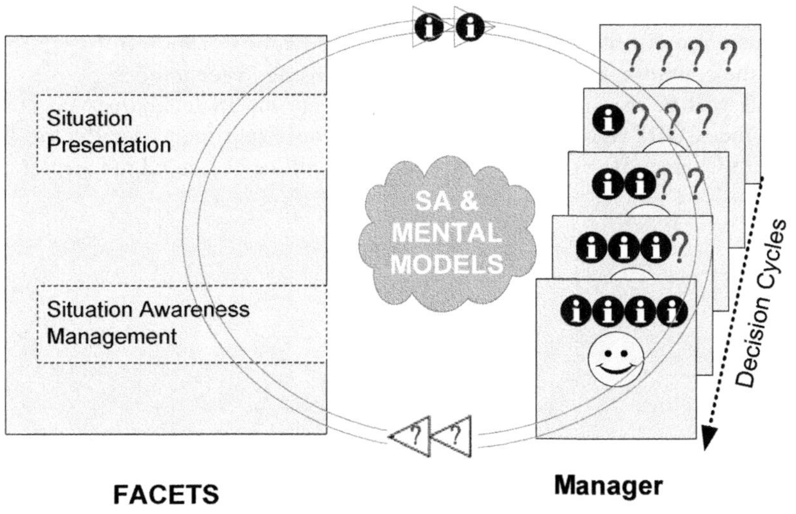

Note:

i --denotes a piece of information/knowledge

?--denotes a question.

Fig. 9.13 The Cognition-Driven Decision Process Based on FACETS

Confronted a decision situation, the decision process a manager would go through is show in Figure 9.13. At the beginning of the decision process, the manager has little SA about the current decision situation (indicated by four question marks). Mangers describe the initial SA in natural language and input SA into FACETS through *Situation Awareness Management.* FACETS analyzes managers' SA and searches for relevant situation information based on the stored domain knowledge: mental models (management experience) and domain ontology. Situation information is then output and presented to the manager for situation assessment. This presentation of situation is also based on managers' mental models (navigation knowledge). Managers develop new SA based on the perceived and understood situation information (indicated by reduced number of question marks and increased number of information marks). The improved SA creates more opportunities for managers to make the right decisions or seek more relevant information. At this point, a decision cycle has been completed according to the CDDP model. This decision cycle can be iterated by the manager until he/she feels confident enough to make reasonable decisions, or he/she is forced to make the final decisions due to limited resources, such as time, money and cognitive load.

9.5 Summary

The prototype system FACETS is reported in this chapter, including its development environment, architecture and functions of subsystems. FACETS is the implementation of the CDDP model. Thus, the major intention of FACETS is to support the cognition-driven decision processes using IS techniques.

FACETS will be used as the test bed to evaluate the IS techniques proposed according the CDDP model. We will present the experiment details for this evaluation in Chapter 10. Three application cases will be presented in Chapter 11.

Chapter 10
Evaluation of Algorithms and FACETS

This chapter reports on the experiments conducted to evaluate the techniques (algorithms) and system (FACETS) developed in this research. Based on a fictitious manufacture company, a data warehouse, an ontology and an experience base were developed and linked with FACETS. Students and professors were invited to the experiments. The FACETS and the major algorithms for SA parsing and query construction were evaluated based on the interaction between subjects and FACETS.

10.1 Experiment Preparation

10.1.1 Data Warehouse

In order to practically evaluate the algorithms, we applied FACETS in an illustrative company, called *Adventure Works Cycles Ltd*, which is a fictitious organization described in the manual of SQL Server 2005 (Microsoft 2007a). *Adventure Works* (AW) is a large, multinational manufacturing company. The company manufactures and sells bicycles and relevant accessories to commercial markets in North America, Europe and Asia. *Adventure Works* is headquartered in Bothell, Washington with 290 employees and several regional sales teams located throughout their market base.

We developed a data warehouse called Adventure Works Data Warehouse (*AWDW*) for this company based on a sample database in SQL Server 2005. In the sample database, there are twenty-nine tables including seven fact tables and twenty-two dimension tables. The data stored in the sample database covers a wide variety of business sectors such as product, account, customer, geography,

L. Niu, J. Lu, and G. Zhang: Cognition-Driven Dec. Supp. for Business Intel., SCI 238, pp. 157–177.
springerlink.com

reseller and sales. In order to make the sample database suitable for this experiment, e.g., creating appropriate business scenarios (decision situations), we extended the sample database in two ways. First, some table definitions were modified. Second, new tables were defined and populated with relevant data. Based on the extended sample database, we developed *AWDW* which represents 75 tables in total with more business sectors covered, such as advertisement, research, government policy, competitors and product delivery. Based on the amended tables, we created six cubes: *internet sales, reseller sales, sales orders, finance, exchange rates* and *delivery*.

10.1.2 Ontology

The data warehouse *AWDW* presents a business application domain. Accordingly, we developed an ontology called Adventure Works Ontology (*AWO*) for this application domain using the *Ontology Management* subsystem of FACETS. In *AWO*, 111 classes are defined which correspond to 111 subsumption relationships in the class tree. Among these classes, 80 classes are correlated to tables or cubes in *AWDW*. We also defined over 1000 property-share relationships across these classes. The excerpt of the class tree and the class graph is shown in Figure 6.1 and Figure 6.2 respectively.

10.1.3 Experience Base

We used the method of experience elicitation discussed in Section 6.2.2 to produce a set of experiences and stored them in the experience base. This experience base is used for all subjects. Experience per se is individual-specific, that is, experience is different from person to person even within the same domain. However, it is also common that people might share the same opinions in some situations, or are able to eventually come up with a consensus after communication. In this sense, it was appropriate to create common experiences for all subjects in this experiment. The advantages of using the same experience base for all subjects are twofold. Firstly, much engineering time in eliciting the experience of every subject was saved. Secondly, the experience base provided a common basis for every subject in the experiments to formulate his/her subjective rating, which might be able to reduce the negative effect of subjective data.

We conducted two experiments to respectively evaluate the related algorithms and the prototype system FACETS. In Experiment one, the following algorithms were evaluated: *NumericPlainParser*, *LiteralPlainParser*, *SemanticParser*, *LocalContextDetermination*, *SqlBuilder* and *MdxBuilder* (Chapter 7). These algorithms represent the major points of the technical part of this research. In Experiment two FACETS, which represents the combination of all related algorithms, was evaluated based on a decision scenario.

10.1.4 Subjects

We invited 22 subjects who have three to five years of working experience in different organizations to conduct the evaluation experiments. Before the experiment, we gave a tutorial to all subjects on the functions and operation of FACETS.

In the experiments, we mainly focused upon the evaluation of relevant IS techniques developed for the situation assessment. The key factors for situation assessment include the usefulness and usability of information presented to the decision maker, such as the quality and quantity of situation information and the method of information presentation.

10.2 Experiment One: Algorithm Evaluation

10.2.1 Experiment Design

We used two basic metrics to measure the performance of algorithms to be evaluated: *precision* and *recall*. Precision and recall are proposed for evaluating the quality of search results returned by information retrieval systems in terms of information relevance . Precision is defined as the fraction of the relevant documents within the collection of all retrieved documents. Recall is the fraction of the documents within the collection of all relevant documents in the data source.

Let ***r*** be the number of relevant documents returned by the system, ***i*** be the number of irrelevant documents returned by the system, and ***m*** be the number of relevant documents in the data source that are not found by the system. The formulas for computing precision and recall are as follows .

$$Precision = \frac{r}{r+i},$$

$$Recall = \frac{r}{r+m}.$$

In this research, the algorithms to be evaluated are the fundamental techniques of situation retrieval. As discussed in Section 5.2, situation retrieval has close relationships with information retrieval. During situation retrieval, different information objects including meta instances (both numeric and literal), classes, native contexts and local contexts are extracted from SA sentences. These information objects can also be judged either relevant or irrelevant to the manager's knowledge need. Therefore, it is reasonable to employ the metrics precision and recall to measure the performance of related algorithms in this research.

In evaluating information retrieval based on precision and recall, a collection of documents needs to be created, and each document needs to be labeled as either relevant or irrelevant. In this experiment, each subject was asked to input 15~20 sentences in English into FACETS. The sentences could be about any topic of the subject's interest within the scope of the *AWO*. For example, subjects could describe products, sales, customer services and government policies, which were defined in the ontology and in the data warehouse. Meanwhile, each subject was asked to specify all the numeric and literal meta instances implied in their SA sentences. By this means, a collection of labeled SA sentences for experiments was created, which was used to evaluate the algorithms *NumericPlainParser* and *LiteralPlainParser*.

Relatively speaking, meta instances are easier to specify due to their simplicity and obviousness. However, SA triples, local contexts and classes, particularly abstract classes, which can be inferred based on meta instances, might be too subtle to be specified by subjects. Thus, in this experiment, subjects were not asked to label these sorts of information, but to judge the outputs of the algorithms by giving 1 (accept) or 0 (reject). Based on the judgments of subjects, precisions of algorithms were calculated. This method was used to evaluate the algorithms *SemanticParser* and *LocalContextDetermination*.

The evaluation of SA triple generation and DW query construction was more complex than the other algorithms, because SA triples and DW queries are very difficult to be explained to subjects, as the end users of the system. However, SA triples, particularly the wording parts, are directly used to construct DW queries which are used to retrieve information from the DW. Thus, analyzing the retrieved information would help us to gain insight into the quality of the underlying DW queries and SA triples. Information evaluation was conducted in Experiment two. In Experiment one, we assessed how parsing length and parsing level affected the average number of SA triples generated. Part of Experiment two can be considered as indirect evaluation of DW query construction and other aspects of SA triple generation, such as information usefulness and information usability.

In Experiment one, we evaluated algorithms *LiteralPlainParser*, *NumericPlainParser*, *SemanticParser* and *LocalContextDetermination*. As these algorithms have already been implemented in FACETS, we set up twenty-two computers with FACETS and relevant databases for the twenty-two subjects. Each subject was able to interact with a FACETS system independently on a computer. The initial data we collected during Experiment one is shown in Table 10.1.

There are 10 columns in Table 10.1. Column *semantic parsing level* is the maximum level at which classes are able to be inferred. *parsing length* is the maximum length of meta instances which can be recognized during plain parsing. The length of a term is the number of words it contains. Experiment one was conducted at four parsing levels (1, 2, 3 and 4) and with different parsing length (1 to 10). For numeric and literal meta instances, metrics *precision* and *recall* are used. For classes and local contexts, *precision* is used. The metric *average number* is also used for classes and SA triples. All other graphs in this section are generated based on Table 10.1.

Table 10.1 Initial Experiment Data of Experiment One

Parsing Level	Parsing Length	Numeric Meta Instance		Literal Meta Instance		Class		Local Context	SA Triple
		p	r	p	r	Average Number	p	p	Average Number
1	1	0.929	0.813	0.731	0.704	2.320	0.753	0.440	1.520
	2	0.929	0.813	0.867	0.963	2.320	0.734	0.640	1.720
	3	0.929	0.813	0.844	1.000	2.360	0.738	0.680	1.760
	4	0.929	0.813	0.871	1.000	2.400	0.741	0.640	1.760
	5	0.929	0.813	0.871	1.000	2.400	0.741	0.640	1.760
	6	0.929	0.813	0.871	1.000	2.400	0.741	0.680	1.760
	7	0.929	0.813	0.844	1.000	2.400	0.750	0.680	1.760
	8	0.929	0.813	0.871	1.000	2.400	0.741	0.640	1.760
	9	0.929	0.813	0.871	1.000	2.400	0.741	0.680	1.760
	10	0.929	0.813	0.871	1.000	2.400	0.741	0.640	1.760
2	1	0.929	0.813	0.731	0.704	3.000	0.789	0.480	2.080
	2	0.929	0.813	0.839	0.963	2.960	0.771	0.680	2.240
	3	0.929	0.813	0.844	1.000	3.040	0.776	0.720	2.280
	4	0.929	0.813	0.844	1.000	3.040	0.776	0.680	2.280
	5	0.929	0.813	0.871	1.000	3.080	0.778	0.720	2.280
	6	0.929	0.813	0.871	1.000	3.080	0.778	0.680	2.280
	7	0.929	0.813	0.871	1.000	3.080	0.778	0.640	2.280
	8	0.929	0.813	0.871	1.000	3.080	0.778	0.680	2.280
	9	0.929	0.813	0.844	1.000	3.040	0.776	0.720	2.280
	10	0.929	0.813	0.844	1.000	3.040	0.776	0.720	2.280
3	1	0.929	0.813	0.704	0.704	3.200	0.792	0.560	2.080
	2	0.929	0.813	0.867	0.963	3.240	0.779	0.720	2.240
	3	0.929	0.813	0.871	1.000	3.280	0.774	0.760	2.280
	4	0.929	0.813	0.844	1.000	3.240	0.771	0.760	2.280
	5	0.929	0.813	0.871	1.000	3.280	0.774	0.760	2.280
	6	0.929	0.813	0.871	1.000	3.280	0.774	0.720	2.280
	7	0.929	0.813	0.844	1.000	3.240	0.771	0.720	2.280
	8	0.929	0.813	0.871	1.000	3.280	0.774	0.760	2.280
	9	0.929	0.813	0.844	1.000	3.240	0.771	0.720	2.280
	10	0.929	0.813	0.844	1.000	3.240	0.771	0.760	2.280

Table 10.1 (*continued*)

4	1	0.929	0.813	0.731	0.704	3.360	0.778	0.560	2.080
	2	0.929	0.813	0.839	0.963	3.320	0.761	0.640	2.240
	3	0.929	0.813	0.871	1.000	3.400	0.766	0.720	2.280
	4	0.929	0.813	0.871	1.000	3.400	0.766	0.680	2.280
	5	0.929	0.813	0.844	1.000	3.360	0.764	0.720	2.280
	6	0.929	0.813	0.871	1.000	3.400	0.766	0.720	2.280
	7	0.929	0.813	0.871	1.000	3.400	0.766	0.680	2.280
	8	0.929	0.813	0.844	1.000	3.360	0.764	0.720	2.280
	9	0.929	0.813	0.844	1.000	3.360	0.764	0.720	2.280
	10	0.929	0.813	0.871	1.000	3.400	0.766	0.680	2.280

Note: *p* denotes precision; *r* denotes recall.

10.2.2 Meta Instance Recognition

Meta instances are recognized during plain parsing. We use two metrics to evaluate the process of meta instance recognition: *precision* and *recall.*

- Numeric Meta Instance Recognition

Numeric meta instances are recognized by algorithm *NumericPlainParser.* As can be seen from Table 10.1, the precision and recall of numeric meta instance recognition remain unchanged across all parsing lengths and parsing levels. Numeric meta instances are mainly fixed-length (one) terms in SA sentences. Thus, the recognition of numeric meta instances is not sensitive to parsing lengths. In the class tree, parsing levels is related to the maximum length of path along which classes are searched, starting from meta instances of taxonomy properties up to higher level abstract classes. Parsing levels do not affect the process of recognizing numeric meta instances.

According to Table 10.1, *NumericPlainParser* is able to recognize numeric meta instances at a precision of 92.9% and a recall of 81.3%.

- Literal Meta Instance Recognition

Literal meta instances are recognized by algorithm *LiteralPlainParser.* The effect of parsing lengths on literal meta instance recognition is shown in Figure 10.1. As each of these four graphs (corresponding to four parsing levels) shows, both precision and recall are affected by parsing length. For example, in Figure 10.1 (a), when parsing length = 1, *LiteralPlainParser* has the lowest precision and recall. With the increase of parsing length from 1 to 2, precision and recall rise quickly from 0.731 to 0.867 and from 0.704 to 0.963 respectively. However, the tendency of

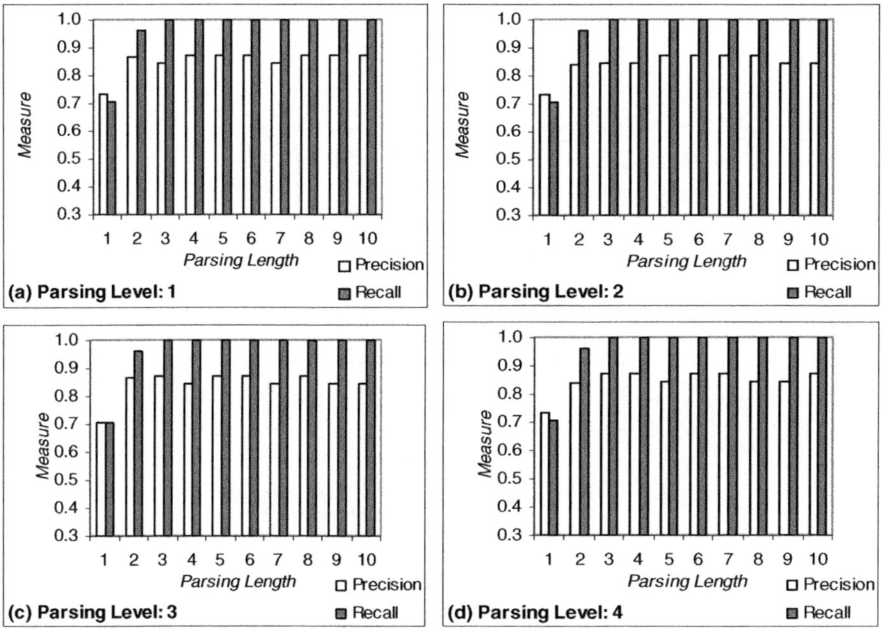

Fig. 10.1 Impact of Parsing Length on Precision and Recall of Literal Meta Instance Recognition

precision and recall to rise quickly does not persist. When parsing length is increased to 3 or greater, recall stays firmly at 100% across the four parsing levels. The turning points of precision are slightly different at different parsing levels. In Figure (a) (parsing level = 1), precision reaches the maximum at parsing length = 4 and then fluctuates slightly around 0.87. For parsing levels 2, 3 and 4, the turning point is respectively at parsing length = 5, 3, and 3.

In SA sentences, different literal instances might have different length. The greater the parsing length, the longer literal instances *LiteralPlainParser* can recognize. For a given parsing length, say ***n***, those literal instances whose length is greater than ***n*** cannot be recognized. This is why precision and recall go up when parsing length increases from 1 to 3. Technically, if parsing length is equal to or greater than the length of the longest literal meta instance in a SA sentence, all literal meta instances can be recognized by *LiteralPlainParser*, which leads to a recall of 100%. Continuous increases of parsing length over the longest literal meta instance, will no longer be able to improve the recall and precision. According to Figure 10.1, it can be inferred that the length of the longest literal meta instance in this experiment is 3, because the recall reaches 100% at a parsing length of 3 or over in four graphs.

Similar to numeric meta instance recognition, Figure 10.2 reflects that parsing level does not significant affect the precision and recall of literal meta instance recognition.

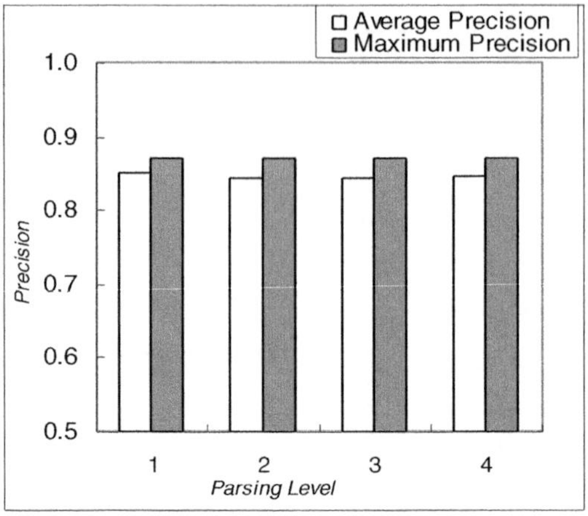

(a) Precision

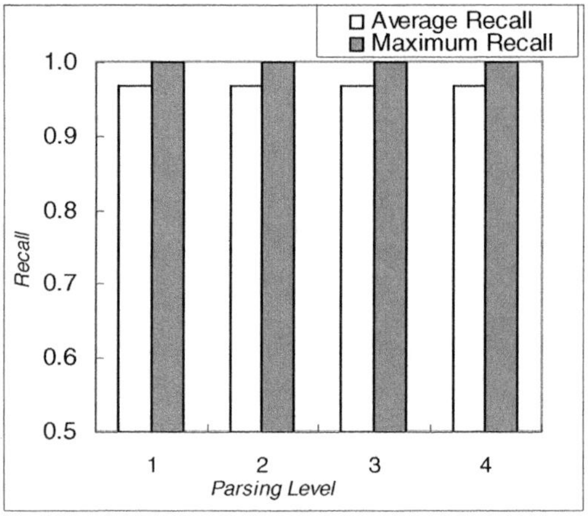

(b) Recall

Fig. 10.2 Impact of Parsing Level on Precision and Recall of Literal Meta Instance Recognition

10.2.3 Class Inferring

Classes are inferred by algorithm *SemanticParser*. We used two metrics to evaluate the process of class inferring: *average number of classes inferred* and *precision*.

(1) Average Number of Classes Inferred

Figure 10.3 shows there is a clear positive relationship between the parsing level and the average of average number of classes inferred: the higher the parsing level, the more classes inferred. The parsing level limits the maximum path length

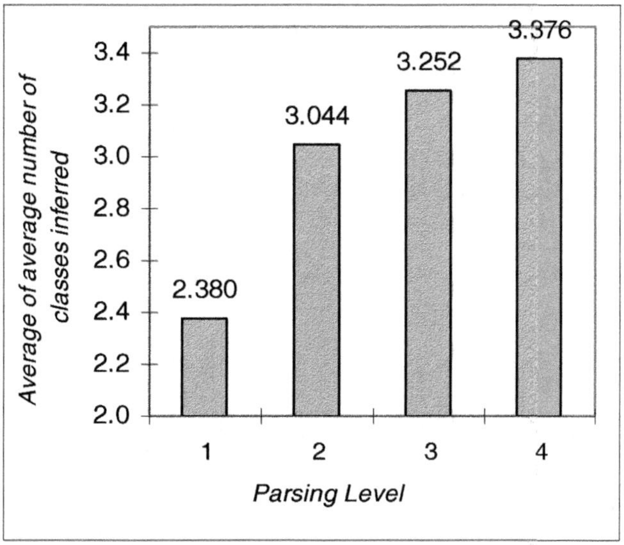

Fig. 10.3 Impact of Parsing Level on Average Number of Classes Inferred

along which algorithm *SemanticParser* searches the class tree. A higher parsing level allows more classes to be searched, which therefore increases the possibility of triggered more appropriate classes.

The impact of parsing length on the average number of classes inferred has different magnitude at different parsing levels (Figure 10.4). The overall pattern demonstrated by the four graphs in Figure 10.4 is that the initial increase of parsing length generally causes a slight increase in the average number of classes inferred. For example, in Figure 10.4 (a), the average number of classes rises from 2.32 to 2.40 when parsing length is increased from 1 to 4. After the initial increase in the average number of classes, no more classes can be inferred with the continuous increase of parsing length. For example, the average number of classes remains 2.400 when parsing length ≥ 4 in Figure 10.4 (a). This pattern becomes unapparent in Figure 10.4 (b), (c) and (d), which correspond to higher parsing levels.

The pattern shown in Figure 10.4 is understandable if we link it to Figure 10.1. According to the algorithm *SemanticParser*, general classes are inferred based on meta instances of their taxonomy properties, which are mainly literal meta instances. The initial increase of parsing length leads to the increase of precision and recall of literal meta instances. In other words, more 'right' literal meta instances are recognized. Consequently, more general classes are triggered by the meta instances. When parsing length rises to 3 or over, the precision and recall of literal meta instance recognition reach the maximum, which also leads to the maximum of the average number of classes.

Exceptional changes are also found in Figure 10.4 (b) and (d). Both graphs show a slight decrease in the average number of classes when parsing length

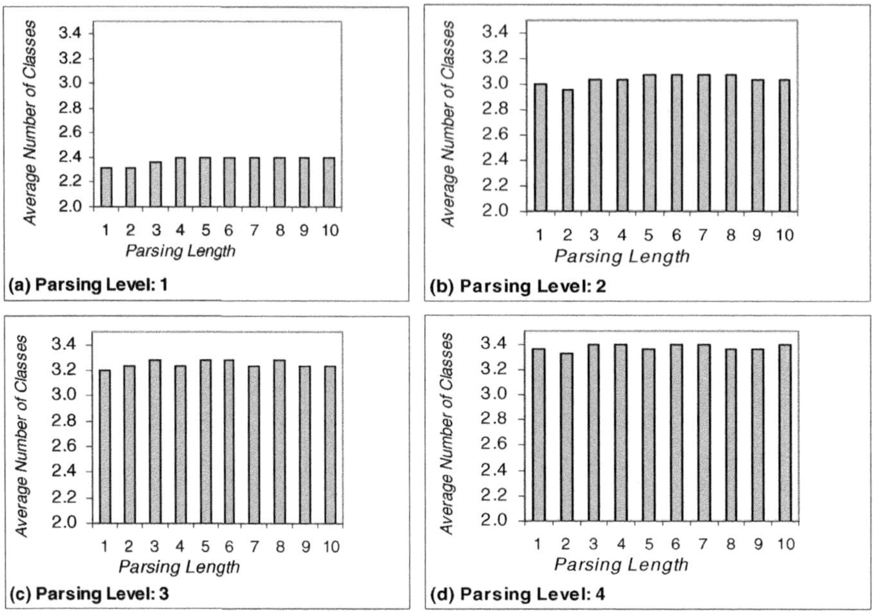

Fig. 10.4 Impact of Parsing Length on Average Number of Classes Inferred

increases from 1 to 2. This might be the impact of other uncontrolled parameters, for example, the subjectivity of experiment data.

(2) Precision of Class Inferring

The precision of class inferring is computed based on subjects' judgments on the output of FACETS. According to Table 10.1, precision of class inferring does not change significantly with the increase of parsing length within each parsing level. The impact of parsing level on precision of class inferring is shown in Figure 10.5. The increase of parsing level from 1 to 2 results in an increase of 0.035 on the precision. However, the precision falls marginally with the continuous increase of the parsing level after 2. The precision change over parsing level is related to the impact of the parsing level on the average number of classes inferred. According to Figure 10.3, the initial increase of parsing level from 1 to 2 results in more classes inferred. The newly inferred 'right' classes out of the all inferred classes make a positive contribution to the improvement of precision. This leads to the initial increase of precision in Figure 10.5. However, the average number of classes inferred keeps going up when the parsing level rises over 2. The higher the parsing level, the more high level classes in the class tree are inferred. Compared to general classes, it is more difficult for subjects to judge if abstract classes, particularly high level abstract classes, produced by the system are right or wrong. This might be the reason why the precision of class inferring changes only slightly over higher parsing levels.

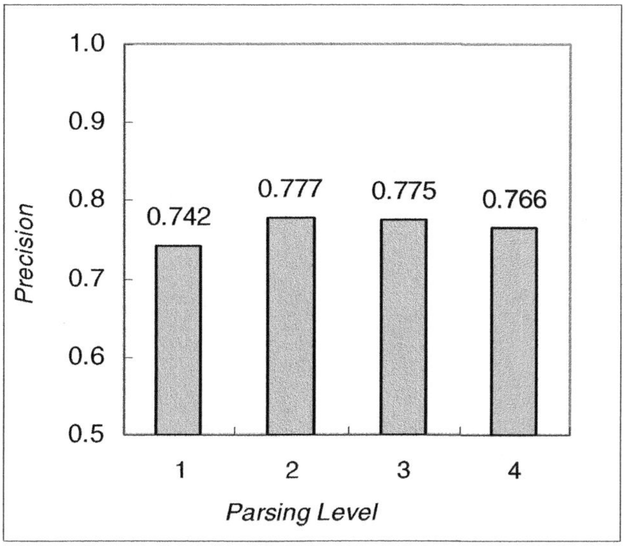

Fig. 10.5 Impact of Parsing Level on Precision of Class Inferring

10.2.4 Local Context Determination

The impact of parsing length on the precision of local context determination is shown in Figure 10.6. With the initial increase of the parsing length from 1 to 3, a rapid increase of precision of local context determination can be seen, followed by fluctuations of precision when parsing length is greater than 3.

Algorithm *LocalContextDetermination* determines the local context for a SA sentence by comparing the context points of alternative classes inferred by algorithm *SemanticParser*. The class with the highest context points will be selected as the local context. Note that there is only one local context for a SA sentence. Thus, as long as the class representing the true local context is inferred, algorithm *LocalContextDetermination* will be likely to find it. According to Figure 10.4, the initial increase of parsing length results in more classes inferred. The increased population of classes will improve the possibility for algorithm *LocalContextDetermination* to find the true local context (in terms of user) out of all inferred classes. As a result, the precision of local context determination goes up with the initial increase of parsing length, and fluctuates with the over increase of parsing length.

Parsing level also has an impact on the precision of local context determination. In Figure 10.7, each average precision is computed at a parsing level across different parsing lengths (from 1 to 10). The corresponding maximum precision is determined by comparing the 10 values of precision. As shown Figure 10.7, both average precision and maximum precision peak at parsing level of 3. This pattern conforms to the impact of parsing level on the precision of classes inferring which is shown in Figure 10.5.

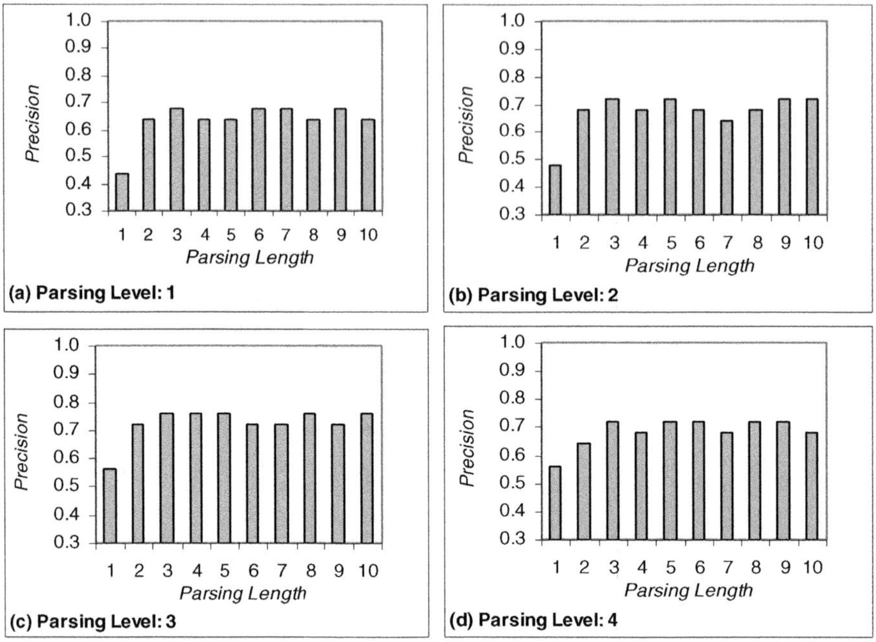

Fig. 10.6 Impact of Parsing Length on Precision of Local Context Determination

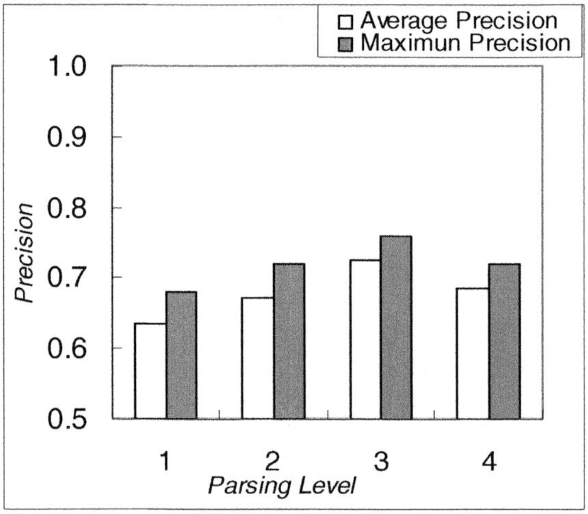

Fig. 10.7 Impact of Parsing Level on Precision of Local Context Determination

The increase of parsing level from 1 to 3 results in more classes inferred which consequently improves the precision of local context determination. However, the over increased number of classes might 'puzzle' the algorithm *LocalContextDetermination,* once the class representing the true local context has been inferred. As a result, the precision of local context determination falls.

10.2.5 SA Triple Generation

The provisional SA triples are generated with uncertainties by algorithms *LiteralPlainParser* and *NumericPlainParser*. The uncertainties are reduced or removed by algorithm *SemanticParser*. Thus, the experiment on SA triple generation reflects the overall performance of these three algorithms. We use metric *average number of SA triples* to evaluate the process of SA triple generation.

Both parsing length and parsing level have a positive impact on the average number of SA triples generated (Figure 10.8). Again, the positive relationship between parsing length and the average number of SA triples lasts only for the initial stage: parsing length rising from 1 to 3. Once parsing length is over 3, no further increase of the average number of SA triples can be seen.

Another noticeable pattern is that, for parsing levels 2, 3, and 4, the average number of SA triples changes in exactly the same way with the increase of parsing length. Comparing the average number of SA triples at each parsing length, a substantial difference exists between parsing level 1 and other three levels.

SA triples generating is based on three algorithms *LiteralPlainParser*, *NumericPlainParser*, and *SemanticParser*. The pattern of SA Triple generation is attributed to the impact combination of parsing length and parsing level on the three algorithms.

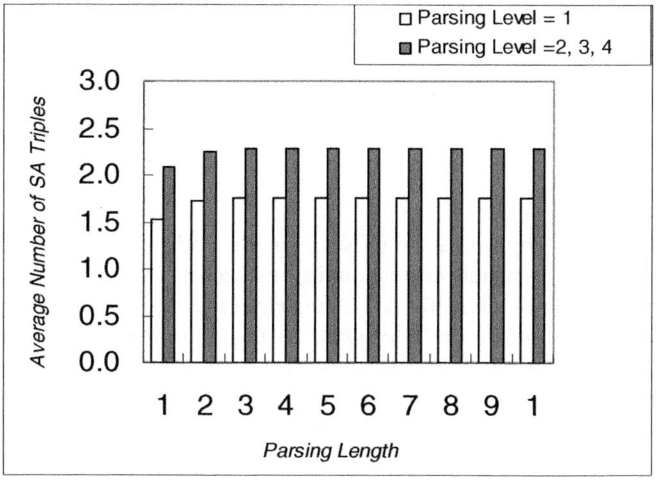

Fig. 10.8 Impact of Parsing Level and Parsing Length on Average Number of SA Triples

10.2.6 Optimization Analysis

Based on the evaluation of algorithms, the optimization values of parsing length and parsing level for SA parsing can be determined (Table 10.2).

Table 10.2 Optimization Parsing Length and Parsing Level

Major Tasks in SA Parsing	Metrics	Optimization Parsing Length	Optimization Parsing Level
Numeric Meta Instance Recognition	Precision	any	any
	Recall	any	any
Literal Meta Instance Recognition	Precision	≥4 , ≥5, **≥3**, ≥3 (respectively for parsing level 1, 2, **3** and 4)	any
	Recall	>=3	any
Class Inferring	Average Number	≥4, ≥5, **≥3**, ≥3 (respectively for parsing level 1, 2, **3** and 4)	4
	Precision	any	2 or 3
Local Context Determination	Precision	**3**	**3**
SA Triple generation	Average Number	≥3	≥2

Note: ***any*** refers to any valid values within 1~4 for parsing level, and 1~10 for parsing length.

In Table 10.2, the optimization parsing lengths and optimization parsing levels refer to the values at which corresponding metrics reach maximum values. The ultimate goal for algorithms *NumericPlainParser, LiteralPlainParser, SemanticParser, LocalContextDetermination* is generating appropriate SA triples and local contexts for query construction. Thus, priority should be given to local context determination and SA triple generation, in determining the optimization values. According to Table 10.2, the optimization parsing level for SA parsing can be roughly determined as 3.

With the determined optimization parsing level, the optimization parsing length for SA parsing can also be identified. Let parsing level = 3, the optimization parsing length for precision of literal meta instance recognition and average number of class inferring is 3. This parsing length (3) agrees to the optimization parsing length for local context determination and SA triple generation (also 3). Thus, the overall optimization parsing length for SA parsing can be roughly determined as 3.

According to Experiment one, for all algorithms, the optimization parsing length = 3 and the optimization parsing level = 3. These optimization values will be used in Experiment two.

10.3 Experiment Two: System Evaluation

10.3.1 Experiment Design

We evaluated FACETS in terms of its usefulness and usability. Usefulness and usability are two dimensions of information value defined by Rouse. The usefulness of a decision support system refers to the extent to which information produced by the system can help users make decisions. The usability is the extent to which information can be easily accessed, understood and applied for decision making. We used a scenario-based experiment to evaluate the usefulness and usability of FACETS.

A decision scenario was created for this experiment based on the ontology, experience base and data warehouse discussed in Section 11.1. The decision scenario set up a decision situation where each subject in the experiment participated in a decision-making process based on FACETS. Subjects described their SA using English and input into FACETS. FACETS called corresponding algorithms to analyze SA, generate DW queries, retrieve relevant situation information and present situation information to the subjects. Perceiving situation information, the subjects updated their SA and re-input into FACETS. In this way, each subject went through up to five decision cycles with the purpose of working out an appropriate decision for the business problem presented in the decision scenario. However, the goal of Experiment two was not to evaluate the outcome of decision generation during a cognition-driven decision process, but to collect subjective data about their decision-making processes. The subjective data was used to evaluate the usefulness and usability of FACETS. During decision cycles, subjects were requested to fill out an evaluation form about the quality of the constructed queries.

During the experiment, each subject ran through up to five decision cycles one by one. During each decision cycle, FACETS generated SQL/MDX reports to present situation information. A SQL report contained data retrieved from a relational table. Subjects browsed the report and judged its usefulness in terms of decision making. A MDX report contained data retrieved from a cube which included a number of dimensions. Subjects also judged the usefulness of dimensions in term of decision making. At the end of each decision cycle, subjects were required to answer three questions:

(1) *How many relational tables/dimensions did FACETS output which you think are useful for your decision situation? The answer is the number of accepted tables/dimensions.*

(2) *How many relational tables/dimensions did FACETS output which you think are useless for your decision situation? The answer is the number of rejected tables/dimensions.*

(3) *How many relational tables/dimensions did FACETS miss (fail to find) which you think are useful for your decision situation? The answer is the number of missed tables/dimensions.*

At the end of the decision process, subjects filled out a survey form to give an overall rating on FACETS. The survey form included twenty-four questions which were designed to elicit subjects' opinions about FACETS as shown in Table 10.4. These questions could be divided into two categories: usefulness evaluation and usability evaluation. Subjects were asked to give their subjective ratings for each question on a scale of five values (*1: strongly disagree, 2: disagree 3: neutral, 4: agree, 5: strongly agree*).

According to the results of Experiment one, the parsing length was set at 3 and the parsing level was set at 3 for all related algorithms in Experiment two.

10.3.2 Query Construction Evaluation

Based on the evaluation forms of query construction that subjects filled out during Experiment two, the performance of query construction was evaluated and is shown in Table 10.3. The precision and recall for each decision cycle were calculated based on the same methods as those in Experiment one.

Table 10.3 Query Construction Performance

Decision Cycles	**1**	**2**	**3**	**4**	**5**
Average # of Accepted Tables/Dimensions	0.9	4.0	8.5	9.2	9.5
Average # of Rejected Tables/Dimensions	1.7	3.1	3.4	3.3	3.6
Average # of Missed Tables/Dimensions	6.7	5.9	4.4	2.1	1.1
Precision	0.35	0.56	0.71	0.74	0.73
Recall	0.12	0.40	0.66	0.81	0.90

Table 10.3 shows that the average number of tables/dimensions accepted by subjects jumps up rapidly from 0.9 to 8.5 over the first three decision cycles. SA is accumulated over decision cycles. The more iterations of decision cycle, the more SA that subjects input into FACETS and consequently FACETS found more relevant situation information. In this sense, the iteration of decision cycles is valuable for decision making. However, the speed of increase in the average

number of accepted tables/dimensions slows after a number of decision cycles. For example, there is only marginal increase of the average number of accepted tables/dimensions during decision cycles 4 and 5. Note that SA was accumulated in FACETS over decision cycles. The information retrieved by FACETS based on SA in a decision cycle might be repeatedly retrieved and presented in a later decision cycle as long as the subject did not change or delete the corresponding SA.

Accompanying the increase of accepted tables/dimensions over decision cycles, the average number of rejected tables/dimensions also goes up, although a slight fall can be seen in decision cycle 4. This means that noise information always comes with valuable situation information in FACETS. In decision cycle 1, there are more tables/dimensions rejected than accepted. After that, accepted tables/dimensions always outnumber the rejected ones. The gap in the quantity of the two sorts of data widens over the decision cycles. This shows that FACETS is able to quickly locate much relevant information and filter noise data for decision making, which can also be seen from the trend of precision change. Precision goes up over the decision cycles. From decision cycle 1 to decision cycle 5, the precision is increased by 0.38.

Over the decision cycles, there are also noticeable changes in the average number of missed tables/dimensions. Less tables/dimensions relevant to decision making were missed by FACETS when it took more SA input. This pattern is also reflected by recall which rises continuously from 0.12 in decision cycle 1 to 0.9 in decision cycle 5.

Based on the above analysis, both the quantity (accepted tables/dimensions) and quality (precision and recall) of information output by FACETS increase significantly over decision cycles. Different stages (decision cycles) of the decision process have various growing speeds: higher in the initial stage, e.g., the first three decision cycles in this experiment, and lower in the latter stage. Possible reasons for the quantity and quality of situation information to stop rising over the decision cycles are (1) newly input SA does not make a significant difference to existing SA; (2) the limitation of knowledge base or the DW is reached. In the case of the reason (1), the decision maker needs to revise the SA description to stimulate FACETS to search for new information. For reason (2), either the knowledge base or the DW needs to be extended by feeding fresh data. The growing speed of information quantity and quality can also act as a type of resource limit discussed in Section 5.2.6, which triggers a final decision to be made. That is to say, if no more new situation information can be found, then probably it is the time to make the final decision.

10.3.3 FACETS Evaluation

Based on the survey forms that subjects filled out during the decision process, the performance of FACETS as a whole was evaluated (Table 10.4). In Table 10.4, the mean rating and standard deviation (SD) for each question (statement) in the survey form were calculated. Questions were grouped into four categories: *situation awareness*, *navigation knowledge map*, *situation information presentation* and

Table 10.4 FACETS Performance

STATEMENTS	Evaluation Category [9]	Mean Rating	Standard Deviation
SITUATION AWARENESS			
1. Given a decision situation, I can precisely describe my SA on it using natural language.	2	4.10	0.67
2. I can easily input my SA into the system.	2	4.92	0.32
NAVIGATION KNOWLEDGE MAPS			
3. Navigation knowledge maps reflect my past thinking (reasoning) processes in decision processes.	1	4.41	0.53
4. Concepts in the navigation knowledge maps are related to my current SA.	1	3.87	0.82
5. Navigation knowledge maps help me to understand the current decision situation.	1	3.35	0.65
6. Navigation knowledge maps have implications for seeking relevant situation information.	1	4.56	0.54
7. Navigation knowledge maps have implications for making decisions in the current decision situation.	1	4.01	0.63
8. I can intuitively make sense of the presentation of a navigation knowledge map (understand its concepts and relationships).	2	4.79	0.51
SITUATION INFORMATION PRESENTATION			
9. The generated reports help me to understand the basic characteristics of the current decision situation.	1	4.68	0.34
10. The generated reports help me to gain in-depth insights into the current decision situation.	1	4.69	0.79

[9] Usefulness:1; Usability: 2.

Table 10.4 (*continued*)

11. The generated reports help me to develop richer SA.	1	3.77	0.81
12. The generated reports help me to seek further situation knowledge.	1	3.01	0.45
13. The generated reports help me to seek further situation information.	1	4.75	0.66
14. The generated report helps me to make final decisions.	1	2.01	0.80
15. The navigation from the presentation of navigation knowledge maps to corresponding reports can be easily followed.	2	4.72	0.30
16. The reports can be easily understood.	2	4.89	0.33
OVERALL			
17. The user interface of FACETS is user friendly.	2	5.00	0.29
18. FACETS helps me to reduce mental workload during the decision process.	1	4.61	1.46
19. FACETS helps me to reuse my past management experience.	1	4.50	0.54
20. FACETS helps me to digest information more easily.	1	3.97	0.42
21. FACETS helps me to obtain valuable information more efficiently.	1	4.03	1.03
22. FACETS helps me to obtain valuable information more effectively.	1	4.67	0.87
23. FACETS helps me to make decisions more rapidly.	1	2.56	0.80
24. FACETS helps me to make decisions more confidently.	1	3.00	0.64

overall. Each question evaluates FACETS from either an information usefulness perspective or from an information usability perspective.

In questions (Q) 1 and 2 of Table 10.4, subjects agreed, using natural language, that they could easily describe their SA and input into FACETS. The natural language interface is important for cognitive decision support. Instead of asking for specific key words in traditional information retrieval systems, FACETS allows managers to describe their SA using natural language. The natural language description of SA can be abstract or very specific. Using natural language, managers can very freely describe and input whatever they think is of significance to their decision making. This might encourage elicitation of valuable information from managers' cognitive processes, which is important to facilitate cognitive decision support.

The overall rating of FACETS in navigation knowledge map evaluation is 4.17. This reflects that the navigation knowledge map is an effective way to present experience for knowledge reuse in decision making. In particularly, navigation knowledge could reflect subjects' past thinking processes (Q3 with average rating of 4.41), and the presentation of navigation knowledge maps is easily to be followed (Q8 with average rating of 4.79). FACETS presents situation information with the guidance of navigation knowledge. Complex decision situations often involve a large quantity of relevant information. Faced with large amount of information, managers are vulnerable to missing their clues without navigation. In FACETS, concepts of interest to decision situations are connected with situation information in navigation knowledge maps. Managers can always re-orientate themselves in the light of the navigation knowledge.

FACETS received an overall rating of 4.07 in situation information presentation evaluation. Subjects agreed that the situation information retrieved by FACETS was of high usefulness in other aspects (Q9 – Q13) and of high usability in Q15 and Q16. However, Subjects in the experiment did not think that the situation information retrieved by FACETS could really help them work out a final decision, for example, Q14 was only scored an average of 2.01, which is the lowest rating in this survey. The possible explanation is that the decision scenario was created based entirely on fictitious business data which lack the sufficient ability to imply real business patterns. Some data might be conflict with the subjects' commonsense and existing knowledge about business.

In the overall evaluation of FACETS, Q17 received the highest rating (5.0): all subjects were fond of the interface of FACETS. Subjects thought their mental workload in decision making were reduced by using FACETS (Q18/4.61). FACETS helped subjects to reuse historical experience in the current decision situation (Q19/4.5). FACETS also improved the efficiency and effectiveness of obtaining valuable situation information (Q21/4.03 and Q22/4.67). Again, FACETS was rated lowly in questions related to actual decision results (Q23 and Q24).

In terms of usefulness, the overall rating of FACETS is 3.91. Thus, the information presented by FACETS is helpful for decision makers to develop relevant SA for decision making. In terms of usability, the overall rating of FACETS is 4.74, which reflects the information generated by FACETS is easy to use for decision making.

FACETS is an implementation of the CDDP model. The performance of FACETS, to some extent, supports our initial expectation on the CDDP model: cognitive decision support. In FACETS, the decision maker's cognitive constructs (mental models and situation awareness) are computerized and represented as information objects, and used to support the process of seeking relevant knowledge and information (situation retrieval) (Q1 and 2). The presentation of the acquired information is also guided by the decision maker's mental models. In this sense, the computerized information processing process is driven by the decision maker's cognition. The ultimate goal of the cognition-driven information processing process is to support the decision maker's cognitive processes for decision making, such as recalling and examining past experience (Q3), perceiving and understanding situation information (Q4, Q5 and Q6), developing situation awareness (Q9, Q10 and Q11), and formulating solutions (Q7) for the current decision situation. In this sense, the argument of cognitive decision support in FACETS is made.

10.4 Summary

This chapter reports the experiments on the major algorithms and the prototype system developed in this research. The key results of the experiments are as follows.

(1) As the support techniques for the CDDP model, the algorithms developed in this research are significantly affected by two factors: parsing length and parsing level. Taking all factors into account, the best parsing length and parsing level are both 3, at which the process of SA parsing reaches the optimization performance.

(2) As the implementation of the CDDP model, the information generated by FACETS has high usefulness and usability. The usefulness evaluation reflects the information generated by FACETS can act as an important utility for decision making. The usability evaluation reflects the information generated by FACETS can be easily accessed, understood and applied for decision making. The evaluation results reflect FACETS is able to support decision making from cognitive aspects.

Chapter 11
Application Cases of FACETS

FACETS was designed and developed using a generic architecture, allowing it to be used for decision support in different domains. We created two fictitious organizations in business (Section 11.1) and public health (Section 11.2) respectively, for illustration of applying FACETS to support decision making. For each specific application of FACETS, three components need to be instantiated: a data warehouse, an ontology, and an experience base.

11.1 Application Case I : Business

11.1.1 Organization Background

Adventure Works (AW) is an international company specialized in manufacturing and selling bikes, bike accessories and related clothing. AW has subsidiaries in Australia, Canada, France, Germany, United Kingdom, and United States. AW markets their products via traditional resellers and internet outlets.

We developed a data warehouse called *AWDW* for this company, based on a sample database in SQL Server 2005. In the sample database, there are twenty-nine tables including seven fact tables and twenty-two dimension tables. The data stored in the sample database covers a wide variety of business sectors such as product, account, customer, geography, reseller and sales. In order to make the sample database suitable for this experiment, e.g., creating appropriate business scenarios (decision situations), we extended the sample database in two ways. First, some table definitions were modified. Second, new tables were defined and populated with relevant data.

AWDW represents 75 tables in total with more business sectors covered, such as advertisement, research, government policy, competitors and product delivery. Based on the amended tables, we created six cubes: *internet sales, reseller sales, sales orders, finance, exchange rates* and *delivery*.

L. Niu, J. Lu, and G. Zhang: Cognition-Driven Dec. Supp. for Business Intel., SCI 238, pp. 179–214.
springerlink.com

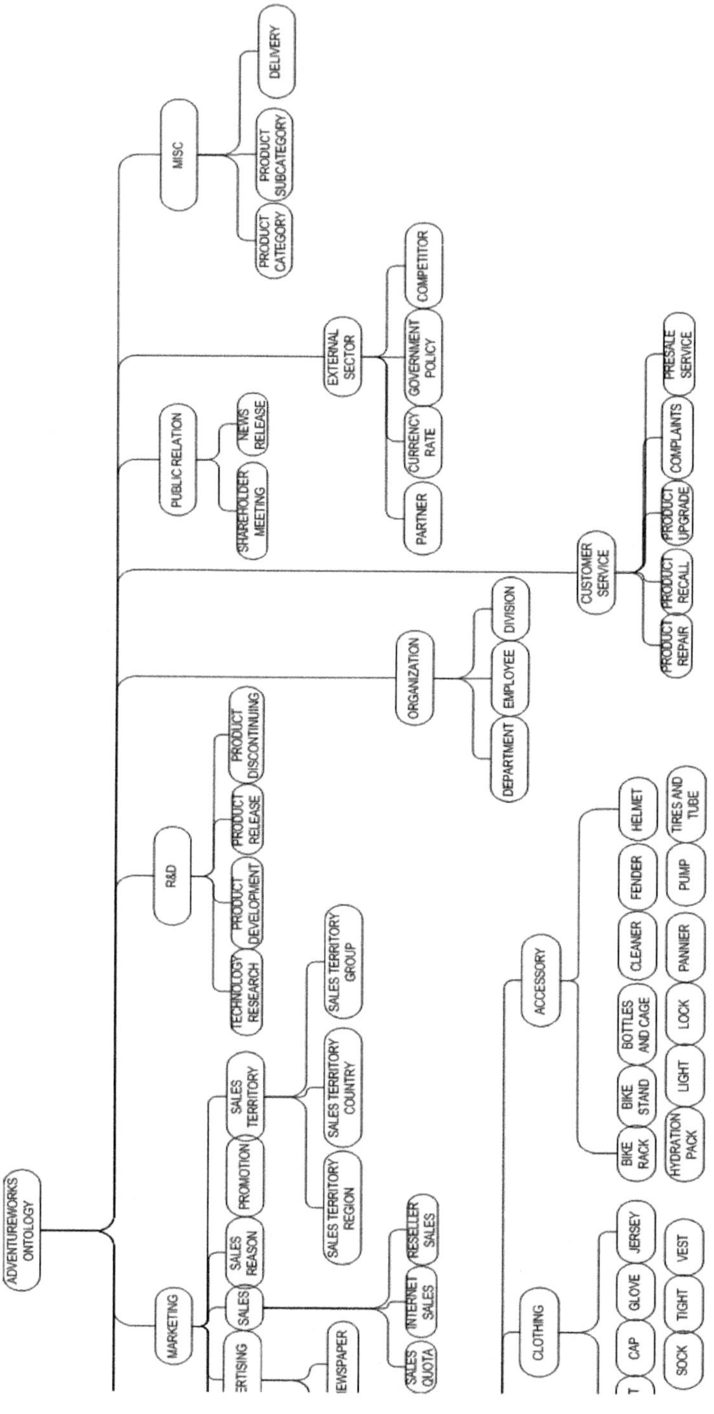

Fig. 11.1 The ClassTree of AWO

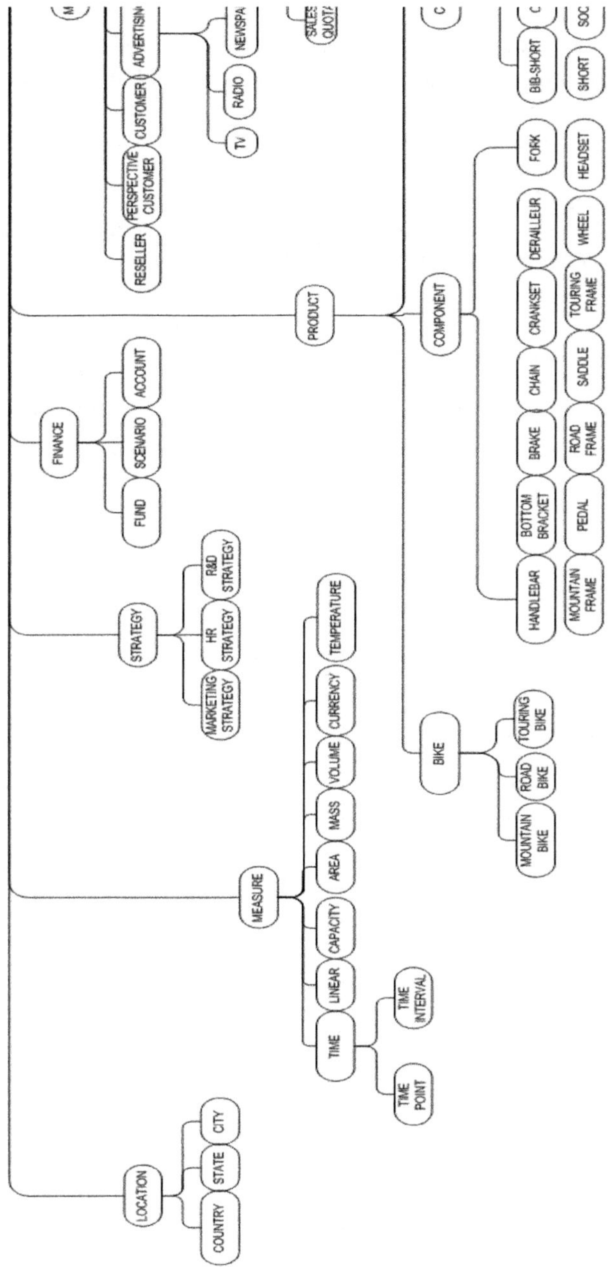

Fig. 11.1 *(continued)*

11.1.2 The Ontology

According to the technical specification of FACETS, we developed an ontology called Adventure Works Ontology (*AWO*) in order to apply FACETS in the context of this company (AW). In *AWO*, totally 116 classes are defined, corresponding to 115 subsumption relationships in the class tree. Among these classes, 80 classes are correlated to tables or cubes in *AWDW*. For example, class PROMOTION has a correlated table *t_promotion* in the *AWDW*. Class INTERNET SALES has a correlated cube *InternetSales* in the *AWDW*. The class tree of *AWO* is shown in Figure 11.1.

We defined over 1000 property-share relationships across the classes in *AWO*. Some examples of property-share relationship are: r^s (PRODUCT RELEASE, SALES), r^s (PRODUCT RELEASE, MOUNTAIN BIKE), and r^s (SALES, PROMOTION).

11.1.3 The Experience Base

The experience base was created using the method of experience elicitation discussed in Section 6.2.2. Part of the experience base is visually shown in Figure 11.2.

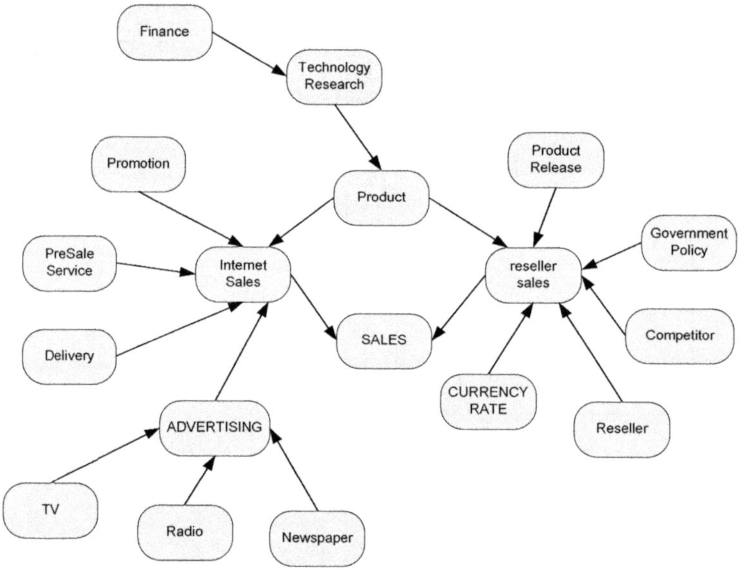

Fig. 11.2 An Excerpt of the Experience Base of Application Case I

The experience map shown in Figure 11.2 represents how PRODUCT SALES are affected by different factors from different business units. This experience map will help a manager to approach the right business information regarding his/her current decision situation. For example, PROMOTION is one of the identified contributors to INTERNET SALES. During a decision process, FACETS might retrieve the data related to the promotion campaigns of BIKES in the market department (stored in the data warehouse), and present the retrieved data to the manager.

11.1.4 Decision Situation

We use a decision situation to illustrate how FACETS can support decision making in the context of AW. This decision situation is described as follows.

> *AW has been dominating the market for over 10 years. However, a big challenge is coming...*
> *Time: 9:00 AM*
> *Date: 24 December 2003*
> *Mr. Cobarol is the chief executive officer of Adventure Works. Mr. Cobarol has been sleepless for days, because he got a very bad news from the marketing department: the sales of their newly released bike model (BK-M82S-38) have dropped over 40% over the past two weeks. So, with FACETS, how should Mr. Cobarol response appropriately, and reverse the tough situation?*

11.1.5 Decision Process

A complete decision process consists of a number of decision cycles. In each decision cycle, the manager inputs into FACETS a description of the current decision situation in form of natural language. The corresponding output of FACETS is the situation information retrieved from the data warehouse according to the situation description. The manager's SA is developed and/or enriched through perceiving the situation information. The improved SA helps the manager to work out a better decision to the current decision situation. There are four decision cycles in the decision process of the decision situation given in Section 11.1.4.

FACETS can be started up through double-clicking the FACETS executable main file under the installation folder. During starting up, FACETS will check all system configuration parameters. If no error is found, the main interface will present to the manager, shown in Figure 11.3.

11.1.5.1 Decision Cycle I

(1) The Initial Situation Awareness

A new decision process begins with inputting the initial SA of the manager.

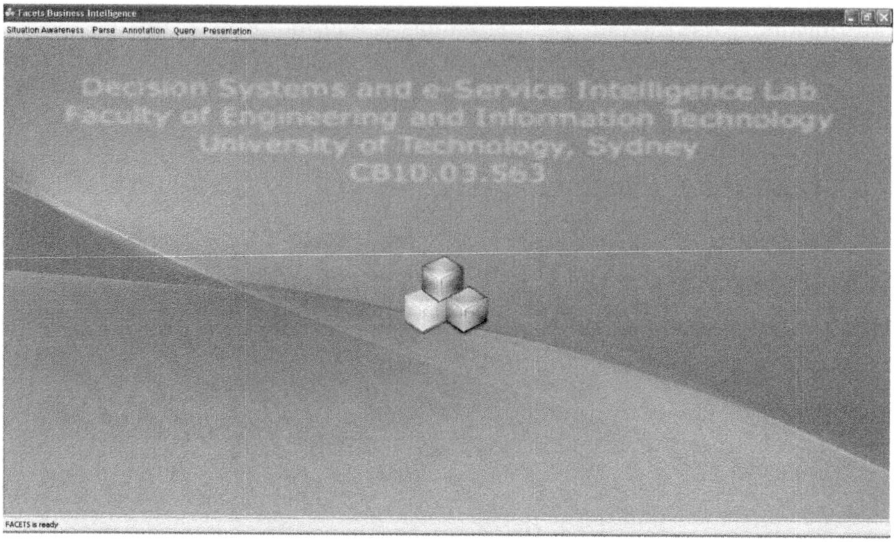

Fig. 11.3 The Main Interface of FACETS

The manager selects the *New Situation* option under menu *Situation Awareness* (Figure 11.4) to open the *Situation Awareness Management* dialog (Figure 11.5).

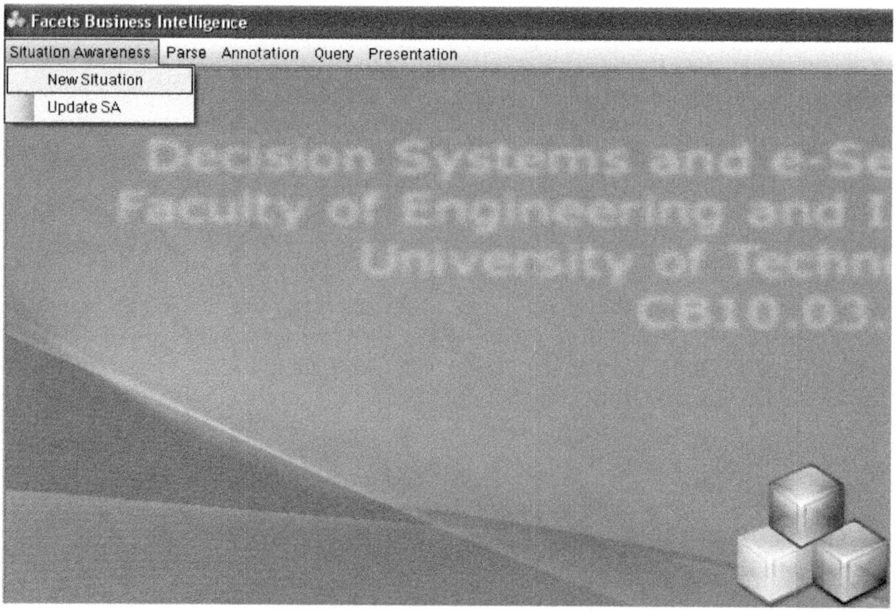

Fig. 11.4 The Menu of Start a New Decision Making Task

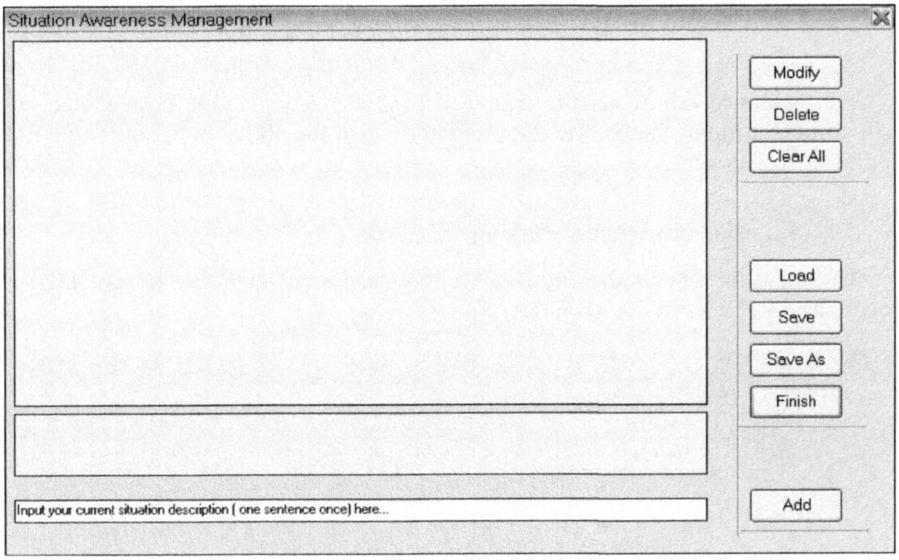

Fig. 11.5 Situation Awareness Management

In the *Situation Awareness Management* dialog, the manager inputs a SA sentence as bellows (Figure 11.6).

Fig. 11.6 The Initial Situation Awareness

BK-M82B-38 was a very good product.

This sentence represents the manager's initial SA about the current decision situation. An initial SA can be more complex than the above one, consisting of a number of sentences.

(2) Situation Awareness Parsing

The manager selects the *Parse SA* option under menu *Parse* (Figure 11.7) to open the *SA Parser* dialog (Figure 11.8).

Fig. 11.7 The Menu of SA Parsing

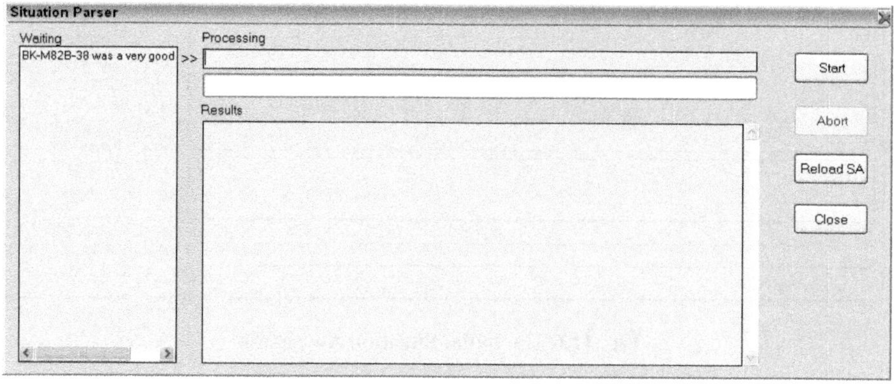

Fig. 11.8 Situation Parser in Decision Cycle I

The manager's initial SA is displayed in the Situation Parser dialog and ready to be parsed by FACETS (Figure 11.8). Figure 11.9 shows the process of SA parsing under progress. Figure 11.10 shows the final results of SA parsing, which are instances or classes inferred by FACETS from the initial SA. The presentation of SA parsing results is mainly for the purpose of research demonstration. As a decision maker, the manager does not need to understand the parsing results.

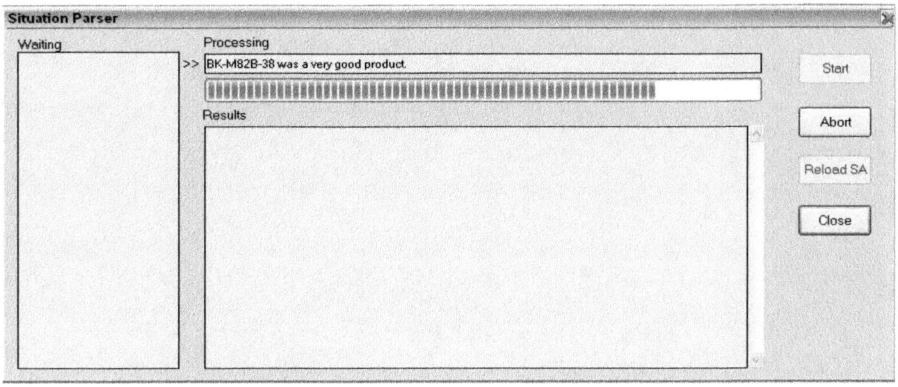

Fig. 11.9 Parse SA Sentences in Decision Cycle I

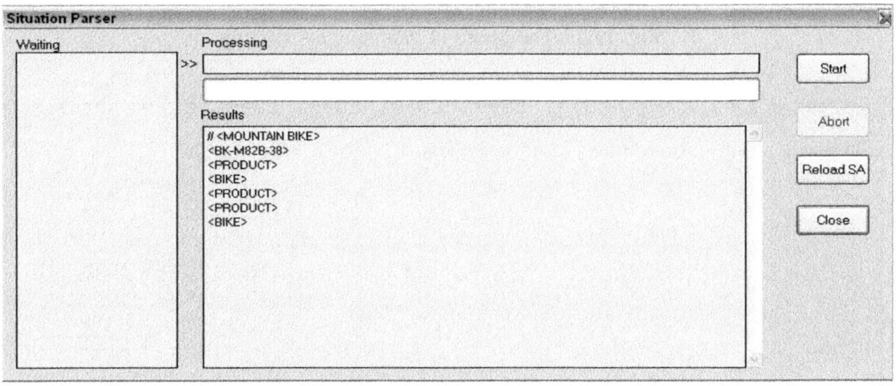

Fig. 11.10 The Results of SA Parsing in Decision Cycle I

(3) Situation Awareness Annotating

The manager selects the *Annotate Current Parse* option under menu *Annotation* (Figure 11.11) to open the *Situation Annotation* dialog (Figure 11.12). The manager clicks on the *Annotate* button, the current SA parsing results are annotated by FACETS, as shown in Figure 11.13. The presentation of SA annotating results is mainly for the sake of research demonstration.

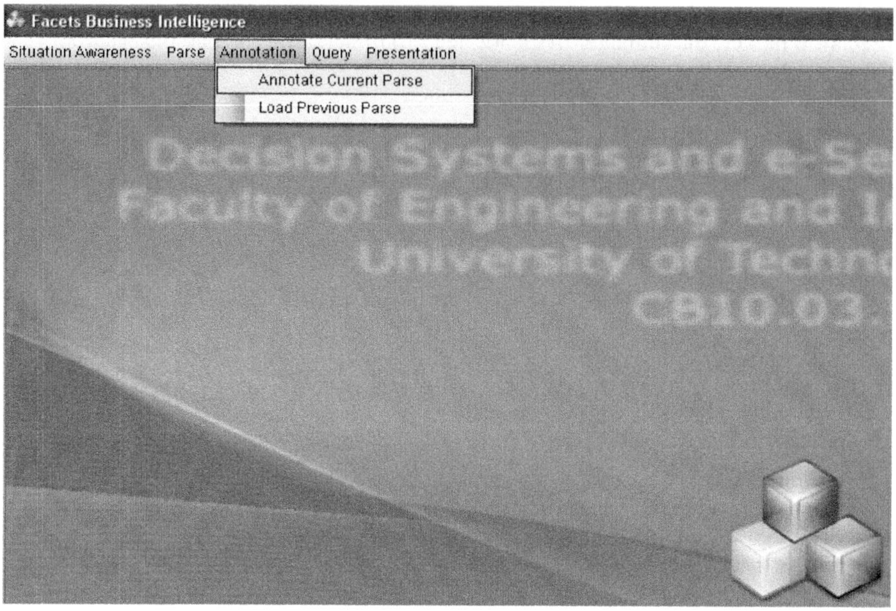

Fig. 11.11 The Menu of Annotate SA Parsing

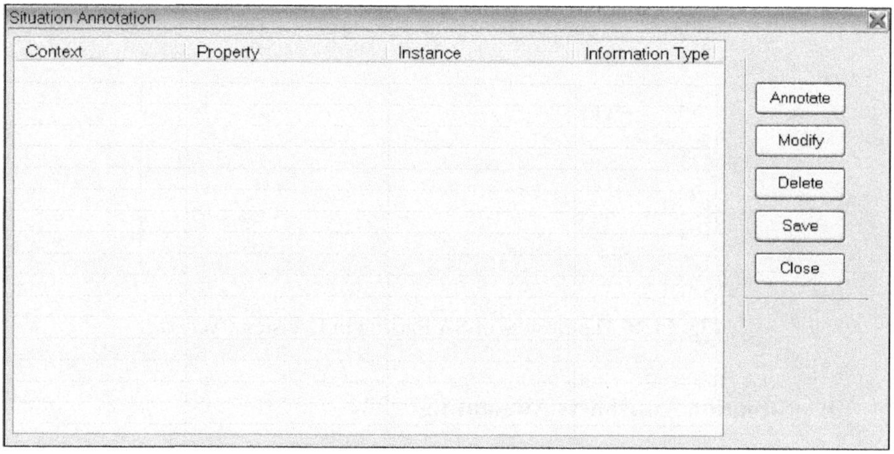

Fig. 11.12 Situation Annotation

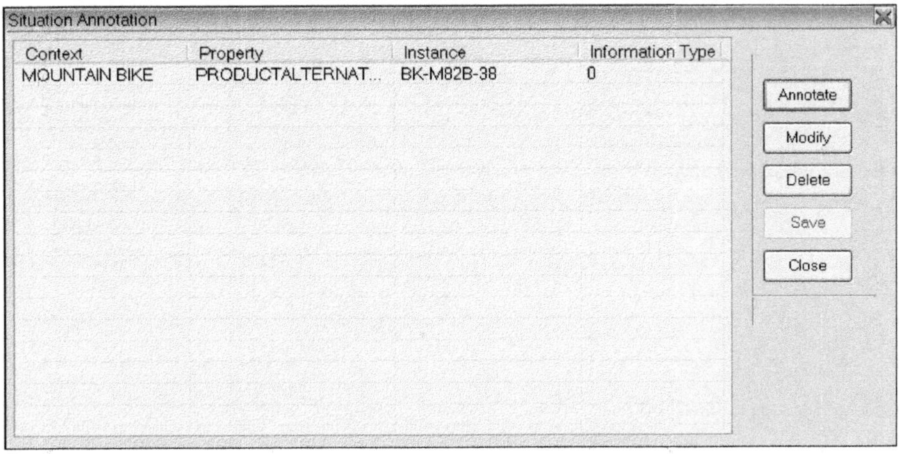

Fig. 11.13 The Results of SA Annotating in Decision Cycle I

(4) Query Generating

The manager selects the *Generate New Query* option under menu *Query* (Figure 11.14) to open the *Query Builder* dialog (Figure 11.15).

The manager clicks on the *Build Queries* button, the corresponding data warehouse queries are generated by FACETS, as shown in Figure 11.16. The presentation of generated queries is mainly for the purpose of research demonstration.

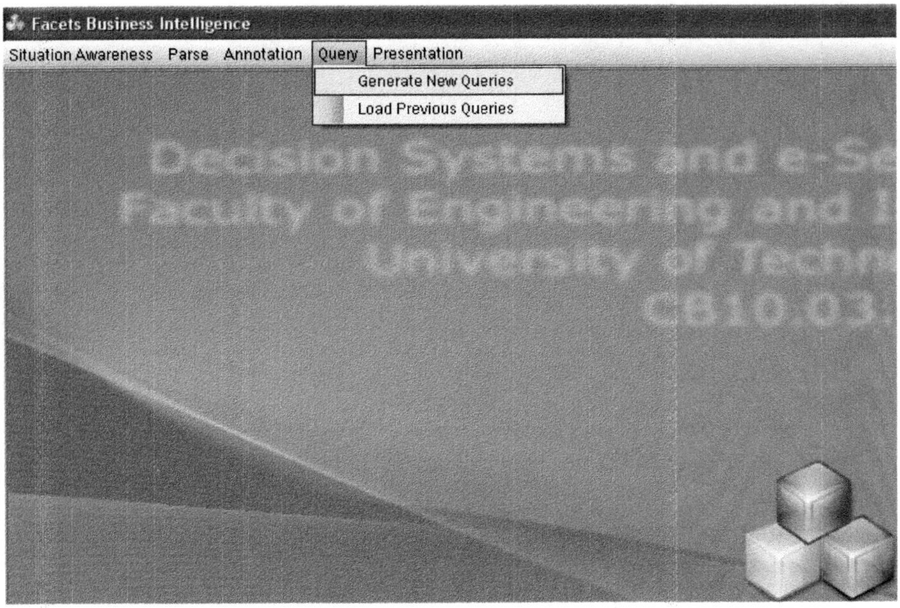

Fig. 11.14 Query Menu

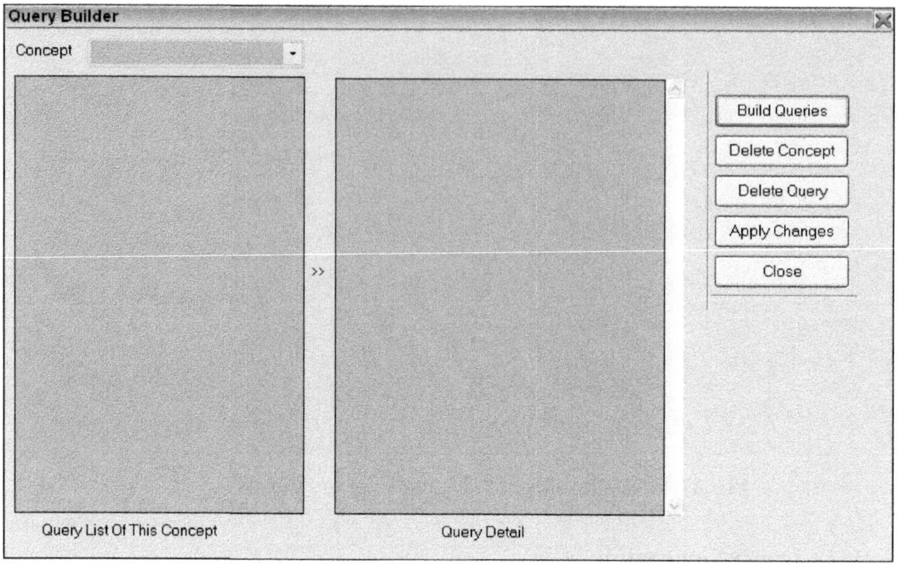

Fig. 11.15 Query Builder

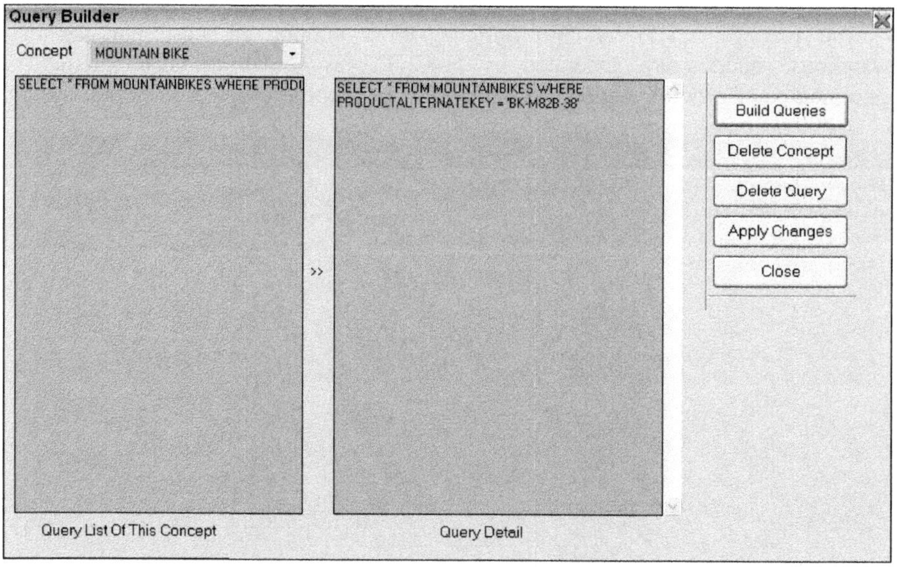

Fig. 11.16 Queries Built by FACETS in Decision Cycle I

(5) Situation Presentation

The manager selects the *Open Cuemap* option under menu *Presentation* (Figure 11.17) to open the corresponding *Cuemap* (Figure 11.18).

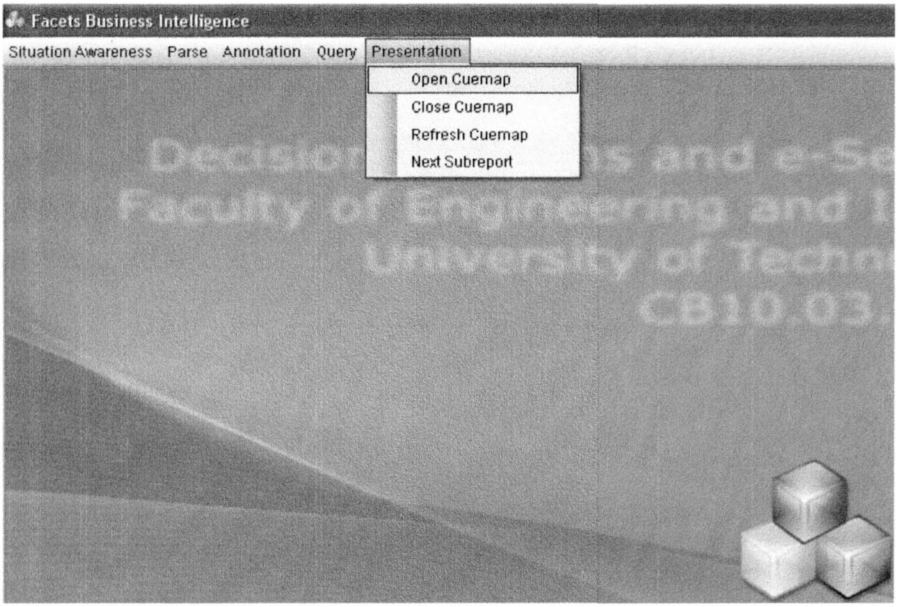

Fig. 11.17 Presentation Menu

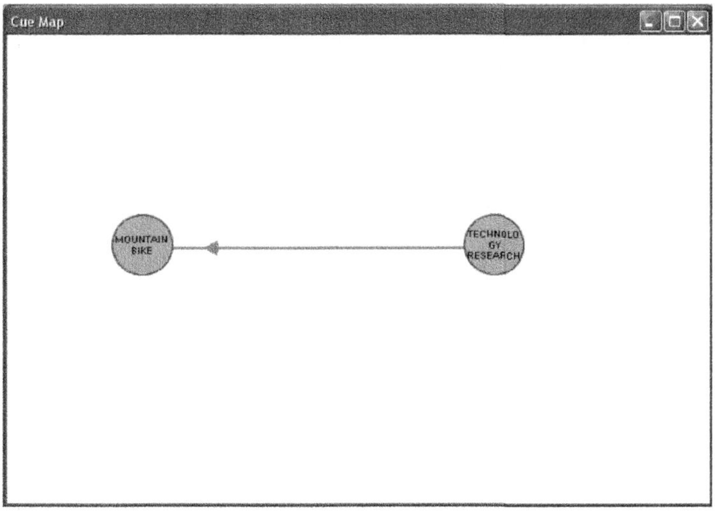

Fig. 11.18 Cuemap in Decision Cycle I

The cuemap is extracted by FACETS from the experience base, according to the manager's input. A node (Figure 11.18) in the cuemap represents a concept of the manager's interest in terms of decision making.

Hovering above a concept icon will activate the corresponding concept. For example, TECHNOLOGY RESEARCH becomes active in Figure 11.19 and MOUNTAIN BIKE active in Figure 11.20.

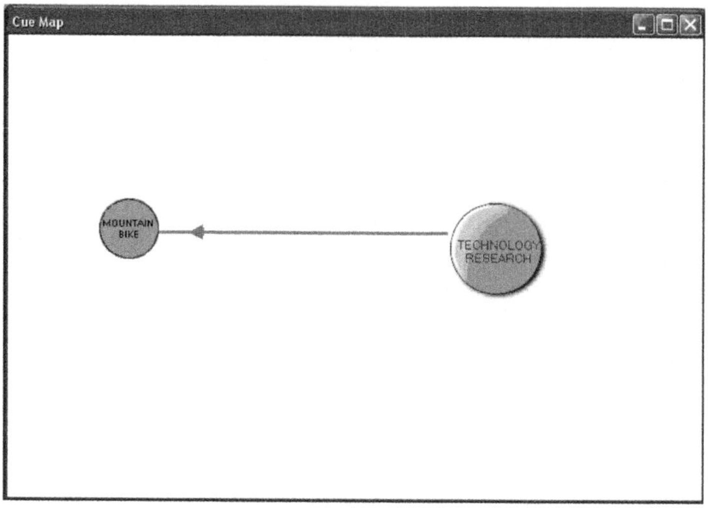

Fig. 11.19 Active Concept: TECHNOLOGY RESEARCH in Decision Cycle I

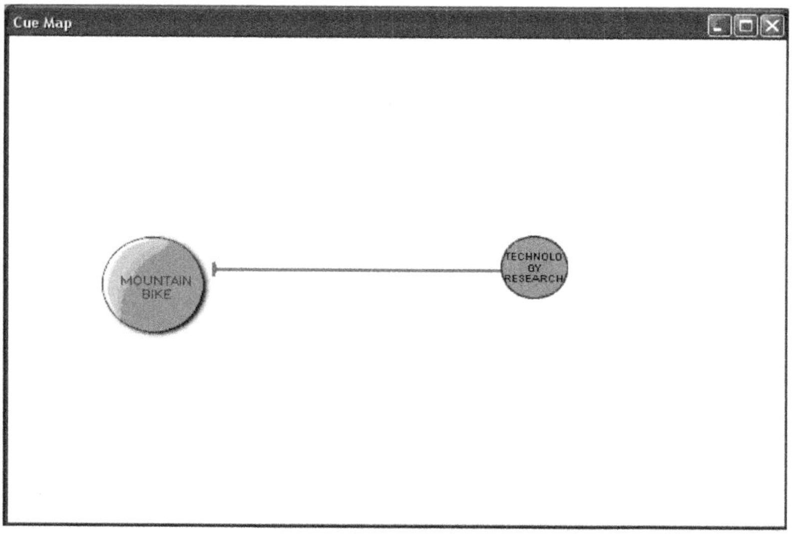

Fig. 11.20 Active Concept: MOUNTAIN BIKES in Decision Cycle I

(6) Reports

Clicking on an active concept will open up the report containing information related to current decision situation. The corresponding report of concept

MOUNTAIN BIKES and TECHNOLOGY RESEARCH in this decision cycle is shown Figure 11.21 and Figure 11.22 respectively.

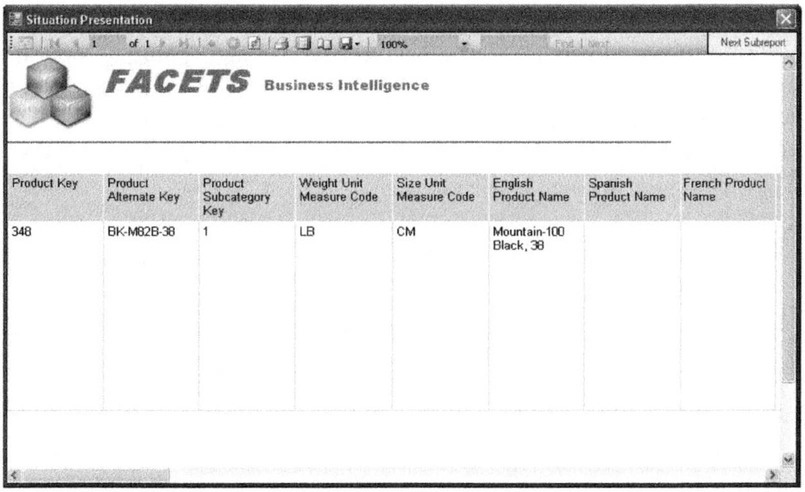

Fig. 11.21 The MOUNTAIN BIKES Report in Decision Cycle I

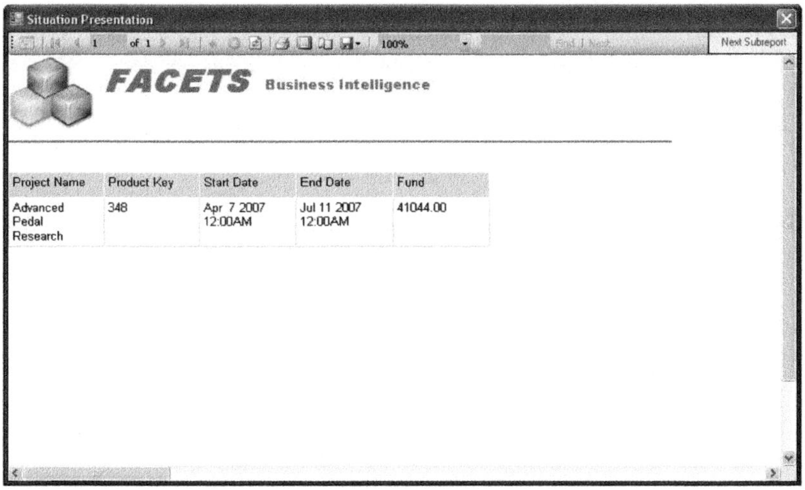

Fig. 11.22 The TECHNOLOGY RESEARCH Report in Decision Cycle I

(7) Situation Assessment

The manager perceives situation information presented in the reports and develops a deeper understanding of the decision situation. As such, the manager's SA is eventually developed and enriched in this decision cycle.

In the end of decision cycle I, the manager found himself not confident enough to make a final decision. Then he chose to proceed to another decision cycle in order to develop richer SA through acquiring more situation information.

11.1.5.2 Decision Cycle II

(1) Situation Awareness Updating

Following the first decision cycle, decision cycle II begins with updating the acquired SA stored in FACETS. The manager selects the *Update SA* option under menu *Situation Awareness* (Figure 11.23) to open the *Situation Awareness Management* dialog (Figure 11.24).

Fig. 11.23 Situation Awareness Menu

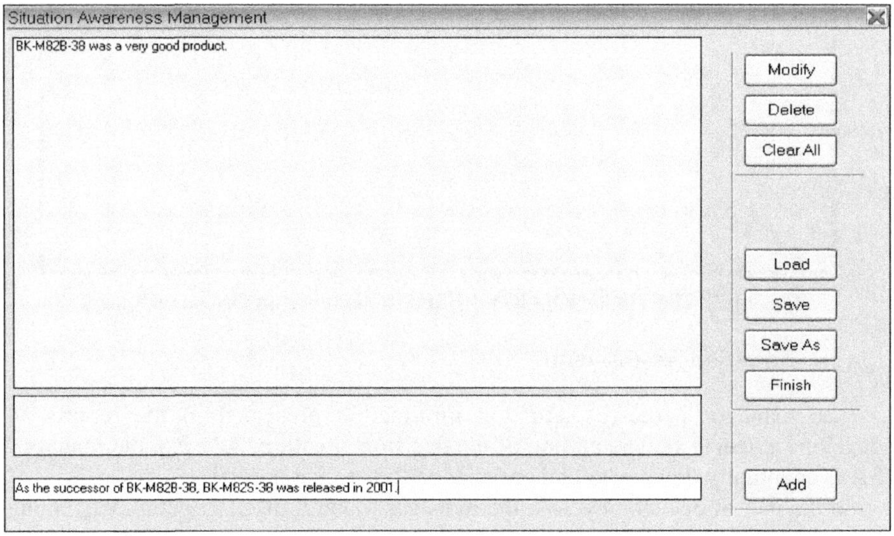

Fig. 11.24 Situation Awareness Update in Decision Cycle II

After reading through the reports presented by FACETS in the first decision cycle, the manager has some updates on his understanding of the decision situation. He updates his SA by inputting the following sentence (Figure 11.24):

As the successor of BK-M82B-38, BK-M82S-38 was released in 2001.

(2) Situation Awareness Parsing

Again, the new SA sentence is parsed by the *Situation Parser* (Figure 11.25).

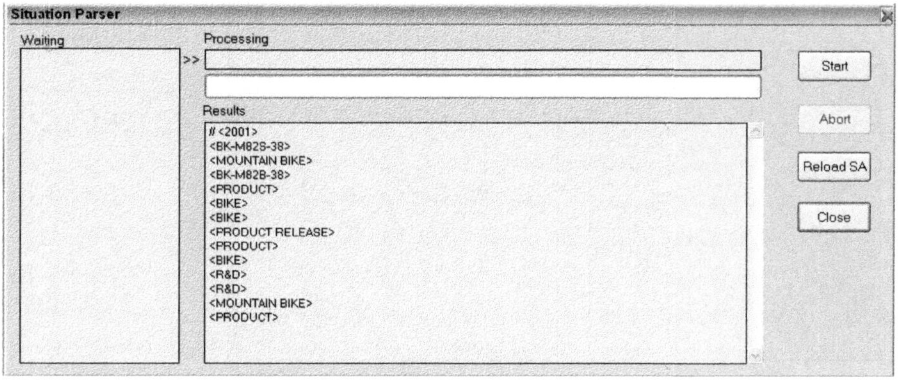

Fig. 11.25 Situation Parser in Decision Cycle II

(3) Situation Awareness Annotating

The manager's new situation awareness is annotated (Figure 11.26).

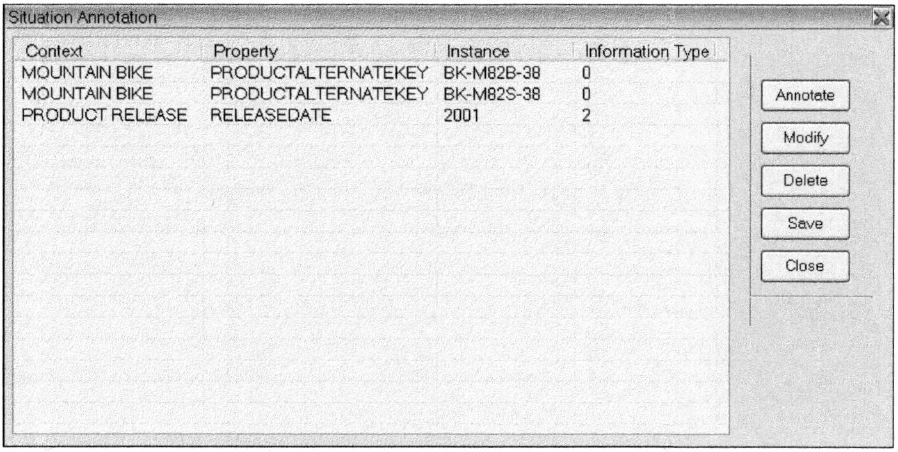

Fig. 11.26 Situation Annotation in Decision Cycle II

(4) Query Generating

The data warehouse queries are re-generated according to the current SA (Figure 11.27 and Figure 11.28).

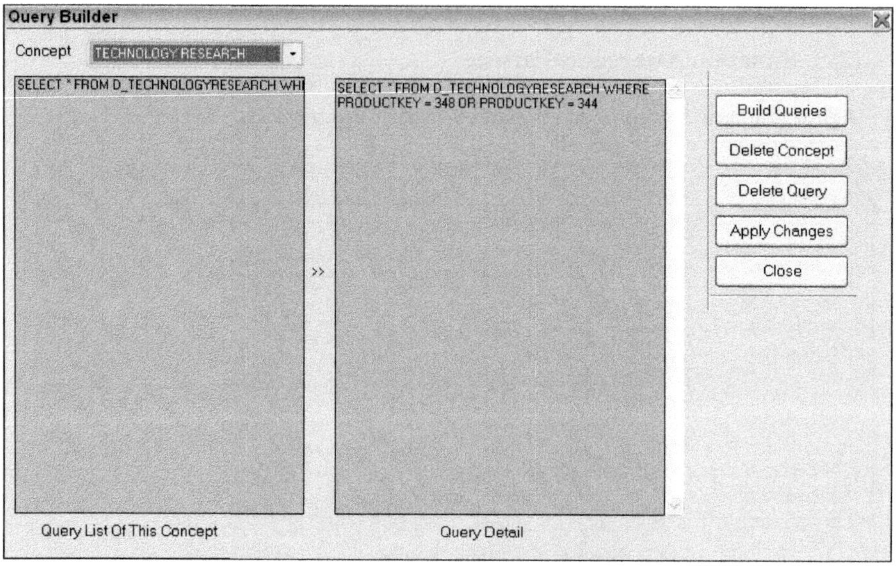

Fig. 11.27 Query Builder in Decision Cycle II

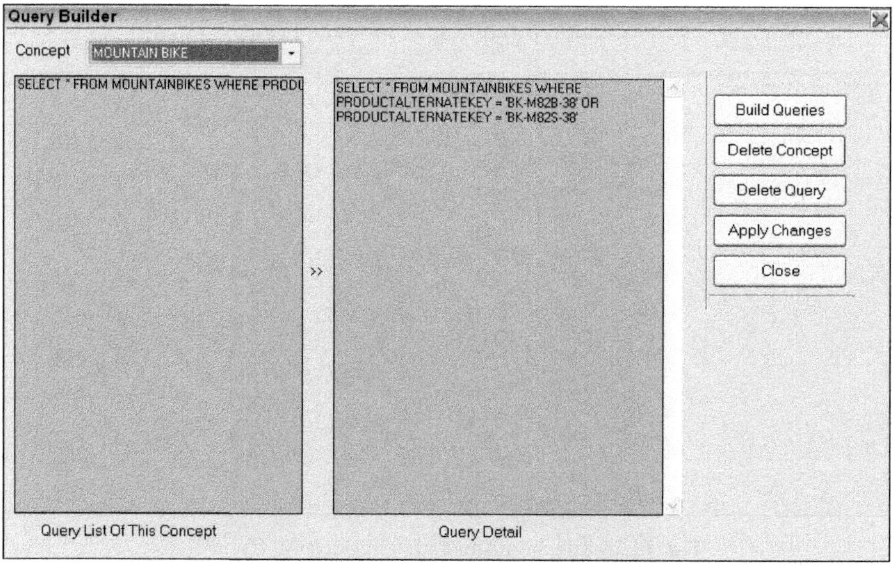

Fig. 11.28 Query Builder in Decision Cycle II

(5) Situation Presentation: Refresh Cuemap

The manager selects the *Refresh Cuemap* option under menu *Presentation* (Figure 11.29) to open the corresponding *Cuemap* (Figure 11.30).

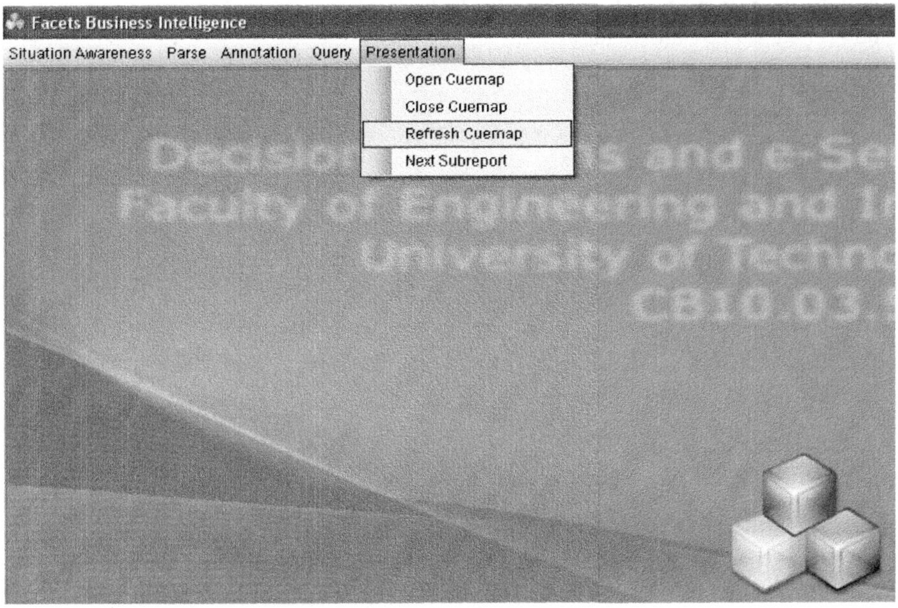

Fig. 11.29 Presentation Menu in Decision Cycle II

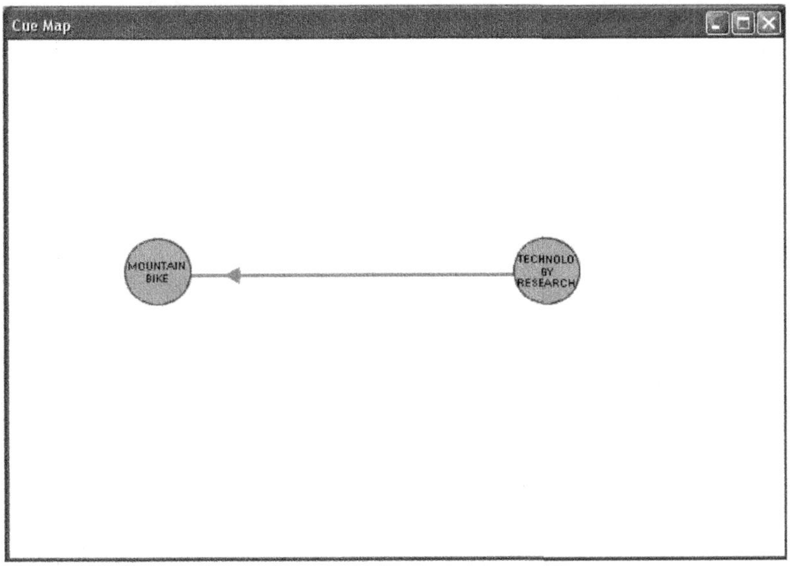

Fig. 11.30 Cuemap in Decision Cycle II

(6) Reports

Clicking on an active concept will open up the report containing information related to current decision situation. The corresponding report of concept MOUNTAIN BIKES and TECHNOLOGY RESEARCH in this decision cycle are shown Figure 11.31 and Figure 11.32 respectively.

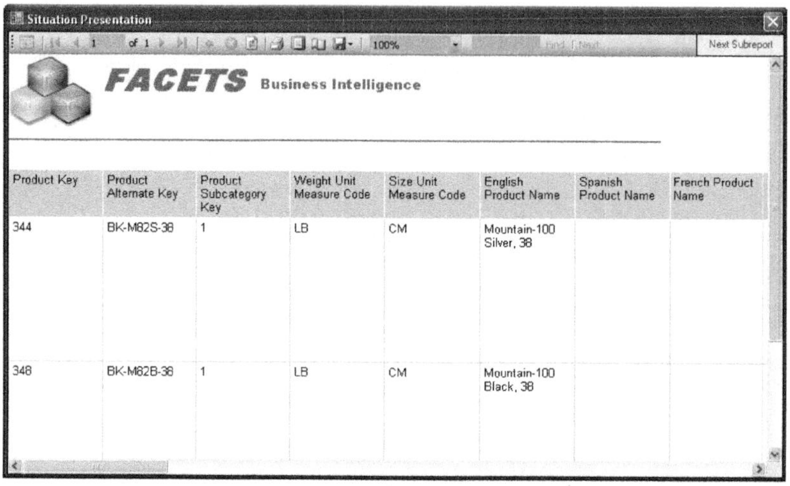

Fig. 11.31 The MOUNTAIN BIKES Report in Decision Cycle Ⅱ

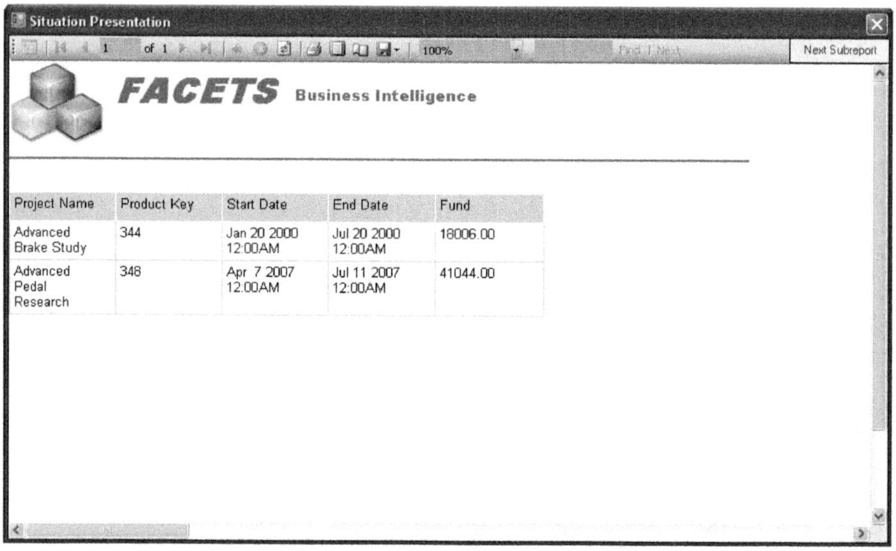

Fig. 11.32 The TECHNOLOGY RESEARCH Report in Decision Cycle Ⅱ

(7) Situation Assessment

The manager perceives situation information presented in the reports and develops a deeper understanding of the decision situation. As such, the manager's SA is eventually developed and enriched in this decision cycle.

In the end of decision cycle Ⅱ, the manager found himself not confident enough to make a final decision. Then he chose to proceed to another decision cycle in order to develop richer SA through acquiring more situation information.

11.1.5.3 Decision Cycle Ⅲ

(1) Situation Awareness Updating

Following the second decision cycle, decision cycle Ⅲ begins with updating the acquired SA stored in FACETS. The manager selects the *Update SA* option under menu *Situation Awareness* (Figure 11.23) to open the *Situation Awareness Management* dialog (Figure 11.24).

After reading through the reports presented by FACETS in the past decision cycles, the manager has some updates on his understanding of the decision situation. He updates his SA by inputting the following sentence (Figure 11.33):

BK-M82S-38 was designed with higher performance and lower price.

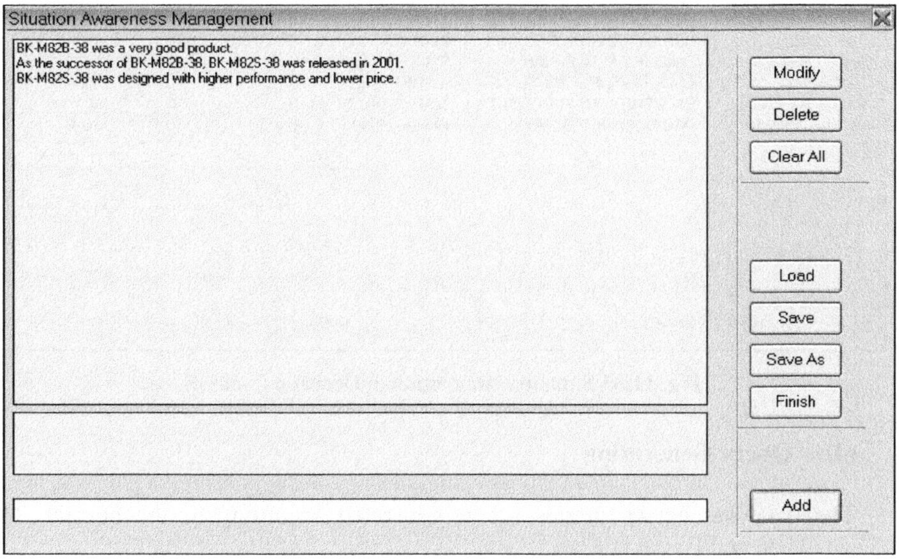

Fig. 11.33 Situation Awareness Update in Decision Cycle Ⅲ

(2) Situation Awareness Parsing

The new SA sentence is parsed by the *Situation Parser* (Figure 11.34).

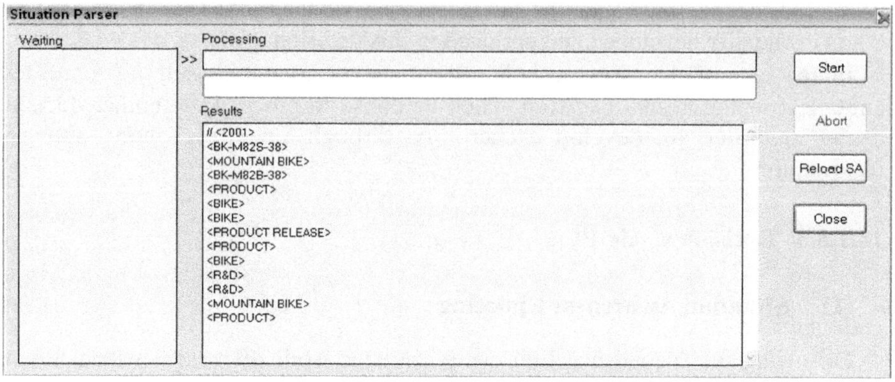

Fig. 11.34 Situation Parser in Decision Cycle Ⅲ

(3) Situation Awareness Annotating

The manager's new situation awareness is annotated (Figure 11.35).

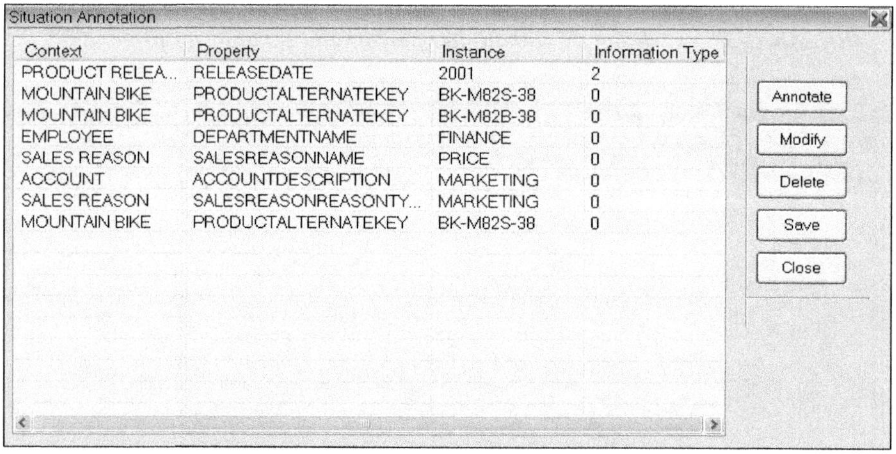

Fig. 11.35 Situation Annotation in Decision Cycle Ⅲ

(4) Query Generating

The data warehouse queries are re-generated according to the current SA (Figure 11.36 and Figure 11.37).

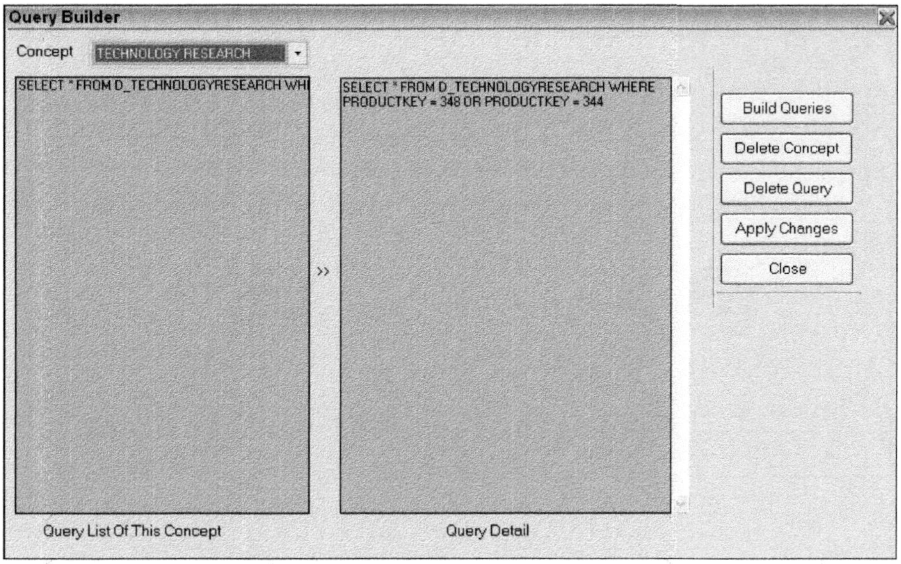

Fig. 11.36 Query Builder in Decision Cycle Ⅲ

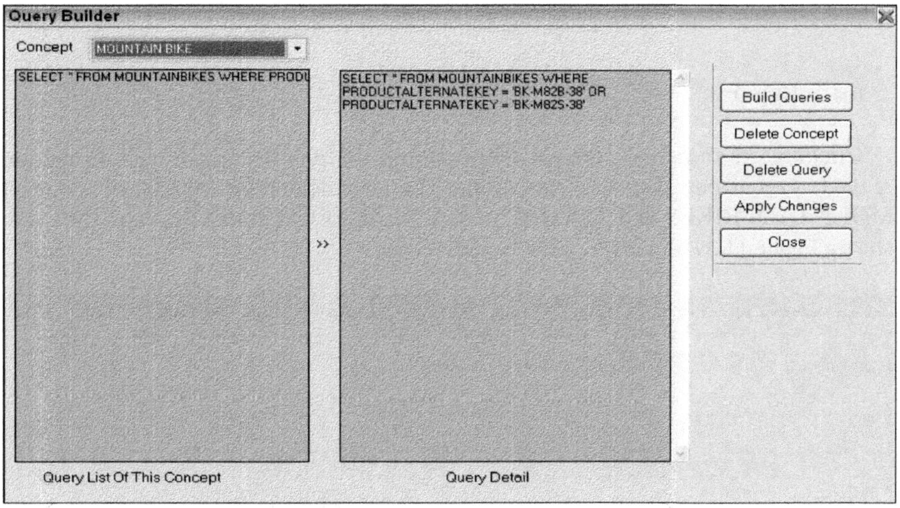

Fig. 11.37 Query Builder in Decision Cycle Ⅲ

(5) Situation Presentation: Refresh Cuemap

The manager selects the *Refresh Cuemap* option under menu *Presentation* to open the corresponding *Cuemap* (Figure 11.38).

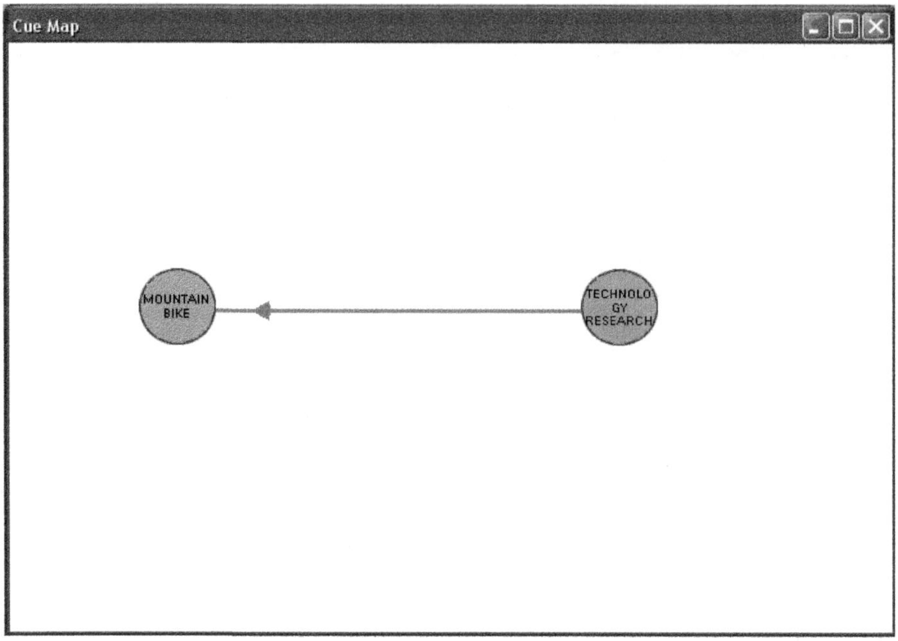

Fig. 11.38 Cuemap in Decision Cycle Ⅲ

(6) Reports

Clicking on an active concept will open up the report containing information related to current decision situation. The corresponding report of concept MOUNTAIN BIKES and TECHNOLOGY RESEARCH in this decision cycle are shown Figure 11.39 and Figure 11.40 respectively.

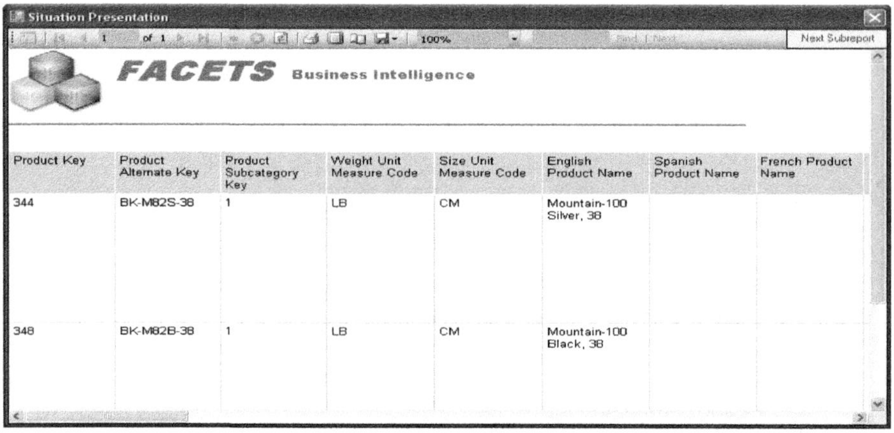

Product Key	Product Alternate Key	Product Subcategory Key	Weight Unit Measure Code	Size Unit Measure Code	English Product Name	Spanish Product Name	French Product Name
344	BK-M82S-38	1	LB	CM	Mountain-100 Silver, 38		
348	BK-M82B-38	1	LB	CM	Mountain-100 Black, 38		

Fig. 11.39 The MOUNTAIN BIKES Report in Decision Cycle Ⅲ

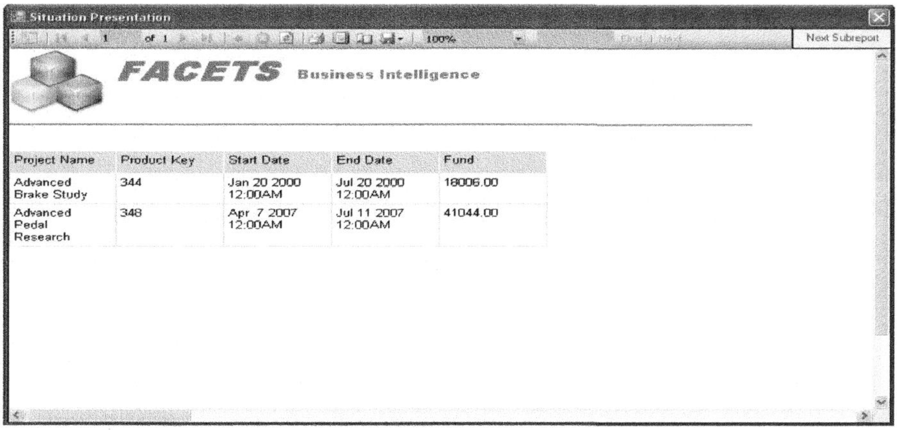

Project Name	Product Key	Start Date	End Date	Fund
Advanced Brake Study	344	Jan 20 2000 12:00AM	Jul 20 2000 12:00AM	18006.00
Advanced Pedal Research	348	Apr 7 2007 12:00AM	Jul 11 2007 12:00AM	41044.00

Fig. 11.40 The TECHNOLOGY RESEARCH Report in Decision Cycle Ⅲ

Within the context of the domain knowledge, FACETS does not detect updates of the manager's input, which is significant enough to retrieve new situation information. Thus, the reports in Decision Cycle Ⅲ are the same as those in Decision Cycle Ⅱ.

(7) Situation Assessment

As there is no any new information related to current decision situation, presented by FACETS in this decision cycle, the manager's SA remains the same as the previous decision cycle.

In the end of decision cycle Ⅱ, the manager found himself not confident enough to make a final decision. Then he chose to proceed to another decision cycle in order to develop richer SA through acquiring more situation information.

11.1.5.4 Decision Cycle IV

(1) Situation Awareness Updating

Following the previous decision cycle, decision cycle IV begins with updating the acquired SA stored in FACETS. The manager selects the *Update SA* option under menu *Situation Awareness* (Figure 11.23) to open the *Situation Awareness Management* dialog (Figure 11.24).

After reading through the reports presented by FACETS in the past decision cycles, the manager has some updates on his understanding of the decision situation. He updates his SA by inputting the following sentence (Figure 11.41):

However, the internet sales of BK-M82S-38 went down in Germany and United Kingdom in 2001 and 2002.

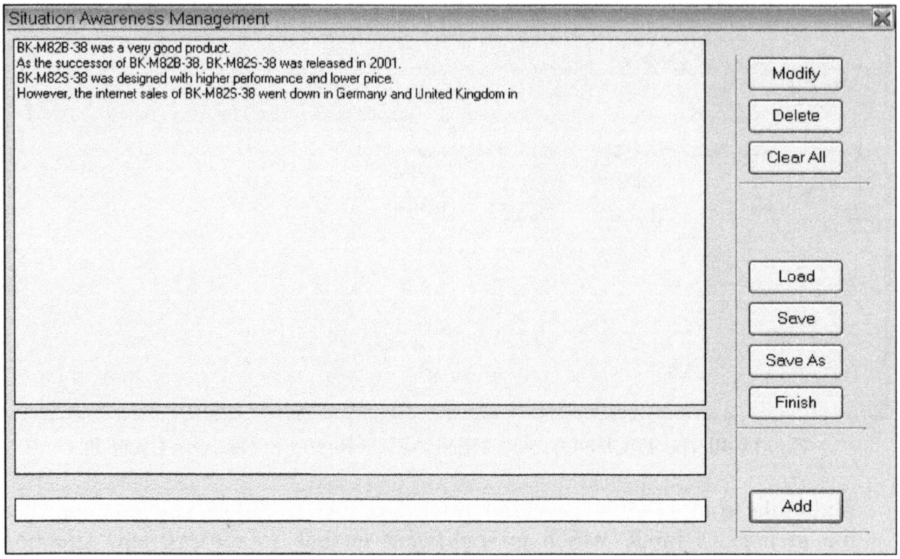

Fig. 11.41 Situation Awareness Update in Decision Cycle IV

(2) Situation Awareness Parsing

The new SA sentence is parsed by the *Situation Parser* (Figure 11.42).

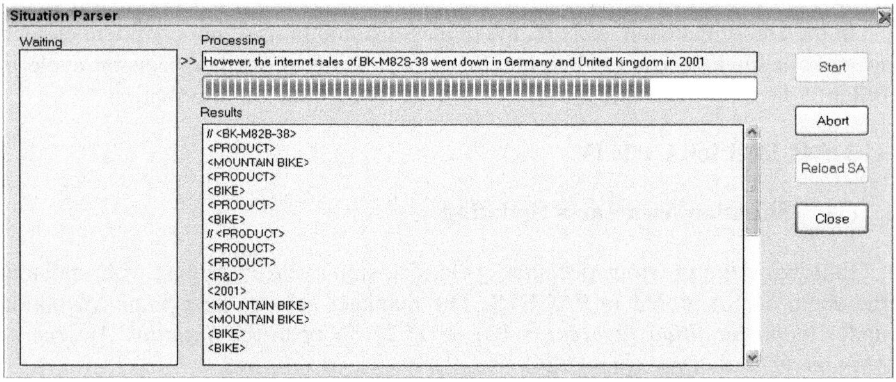

Fig. 11.42 Situation Parser in Decision Cycle IV

(3) Situation Awareness Annotating

The manager's new situation awareness is annotated (Figure 11.43).

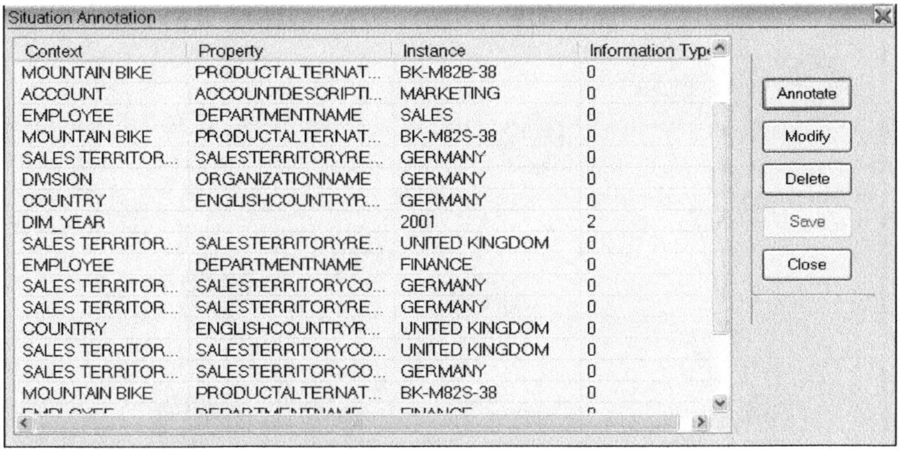

Fig. 11.43 Situation Annotation in Decision Cycle IV

(4) Query Generating

The data warehouse queries are re-generated according to the current SA (Figure 11.44).

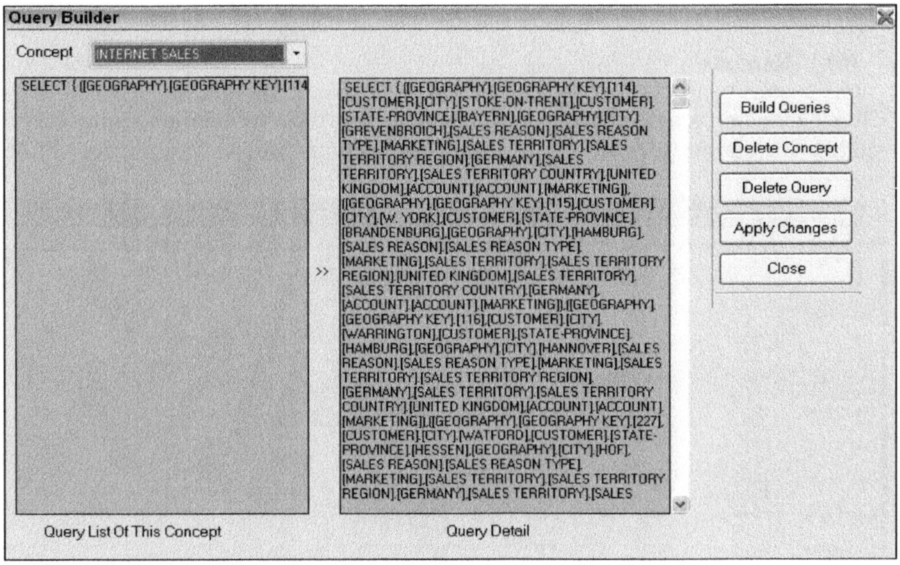

Fig. 11.44 Query Builder in Decision Cycle IV

(5) Situation Presentation: Refresh Cuemap

The manager selects the *Refresh Cuemap* option under menu *Presentation* to open the corresponding *Cuemap* (Figure 11.45).

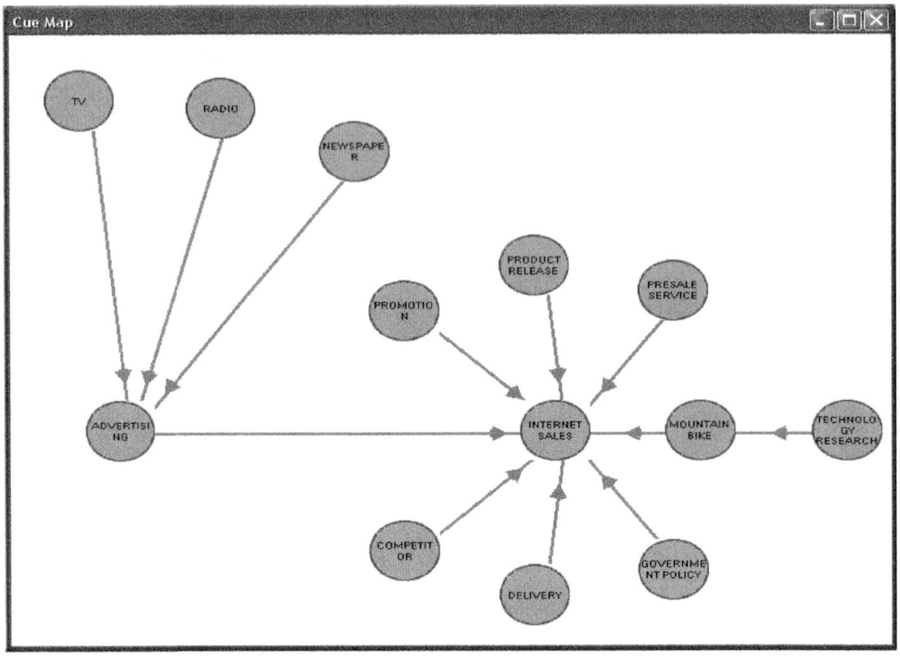

Fig. 11.45 Cuemap in Decision Cycle IV

(6) Reports

Clicking on an active concept will open up the report containing information related to current decision situation. The reports shown in Figure 11.46,

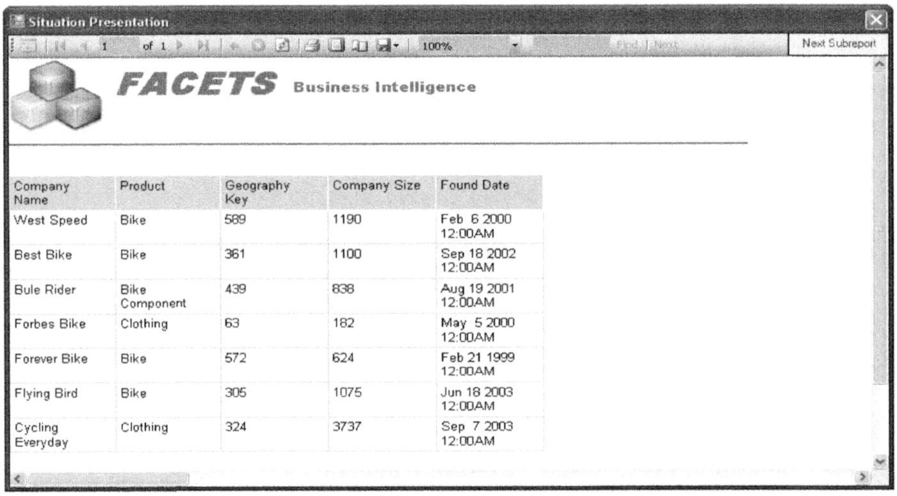

Company Name	Product	Geography Key	Company Size	Found Date
West Speed	Bike	589	1190	Feb 6 2000 12:00AM
Best Bike	Bike	361	1100	Sep 18 2002 12:00AM
Bule Rider	Bike Component	439	838	Aug 19 2001 12:00AM
Forbes Bike	Clothing	63	182	May 5 2000 12:00AM
Forever Bike	Bike	572	624	Feb 21 1999 12:00AM
Flying Bird	Bike	305	1075	Jun 18 2003 12:00AM
Cycling Everyday	Clothing	324	3737	Sep 7 2003 12:00AM

Fig. 11.46 The COMPETITOR Report in Decision Cycle IV

Figure 11.47 and Figure 11.48 are respectively connected to concepts COMPETITOR, PROMOTION, and DELIVERY.

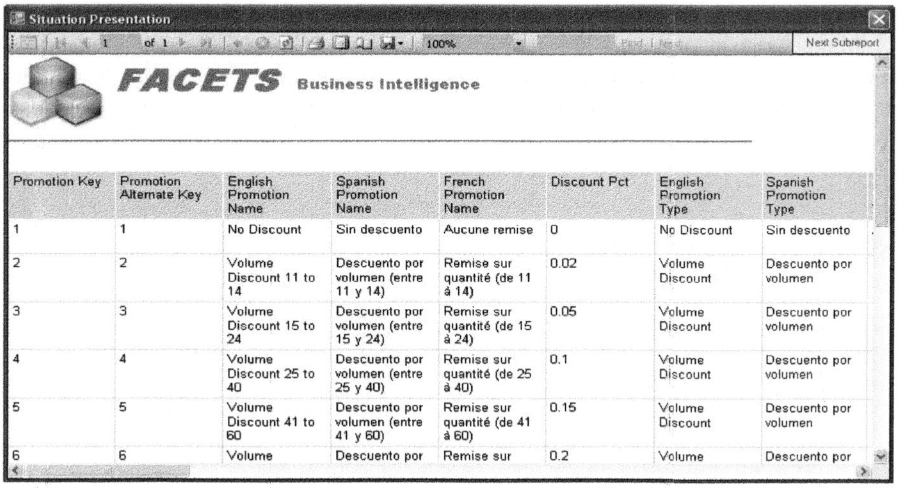

Promotion Key	Promotion Alternate Key	English Promotion Name	Spanish Promotion Name	French Promotion Name	Discount Pct	English Promotion Type	Spanish Promotion Type
1	1	No Discount	Sin descuento	Aucune remise	0	No Discount	Sin descuento
2	2	Volume Discount 11 to 14	Descuento por volumen (entre 11 y 14)	Remise sur quantité (de 11 à 14)	0.02	Volume Discount	Descuento por volumen
3	3	Volume Discount 15 to 24	Descuento por volumen (entre 15 y 24)	Remise sur quantité (de 15 à 24)	0.05	Volume Discount	Descuento por volumen
4	4	Volume Discount 25 to 40	Descuento por volumen (entre 25 y 40)	Remise sur quantité (de 25 à 40)	0.1	Volume Discount	Descuento por volumen
5	5	Volume Discount 41 to 60	Descuento por volumen (entre 41 y 60)	Remise sur quantité (de 41 à 60)	0.15	Volume Discount	Descuento por volumen
6	6	Volume	Descuento por	Remise sur	0.2	Volume	Descuento por

Fig. 11.47 The PROMOTIION Report in Decision Cycle IV

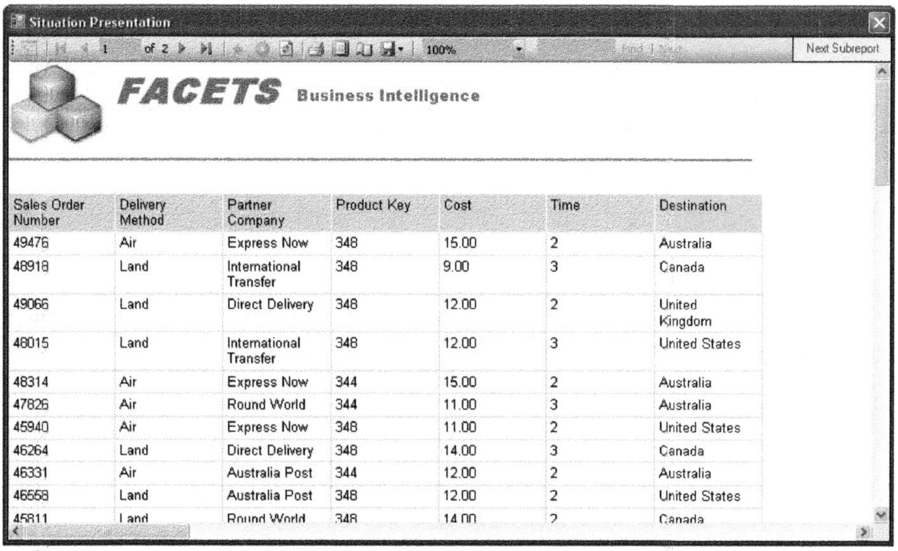

Sales Order Number	Delivery Method	Partner Company	Product Key	Cost	Time	Destination
49476	Air	Express Now	348	15.00	2	Australia
48918	Land	International Transfer	348	9.00	3	Canada
49066	Land	Direct Delivery	348	12.00	2	United Kingdom
48015	Land	International Transfer	348	12.00	3	United States
48314	Air	Express Now	344	15.00	2	Australia
47826	Air	Round World	344	11.00	3	Australia
45940	Air	Express Now	348	11.00	2	United States
46264	Land	Direct Delivery	348	14.00	3	Canada
46331	Air	Australia Post	344	12.00	2	Australia
46558	Land	Australia Post	348	12.00	2	United States
45811	Land	Round World	348	14.00	2	Canada

Fig. 11.48 The DELIVERY Report in Decision Cycle IV

If the concept being clicked corresponds to a cube, the *MDX reporting* module will be called. The MDX reporting module requires managers to choose firstly the analysis condition. Analysis conditions consist of dimensions, attributes, members and measures. As shown in Figure 11.49, a number of relevant dimensions such as RESELLER, SALES TERRITORY, PRODUCT and DATE are available for selection. Under each dimension are a number of attributes; under each attribute are a number of members. For example, Sales Territory Country is an attribute for dimension SALES TERRITORY; United Kingdom and Germany are members for attribute Sales Territory Country.

Managers can organize the analysis conditions of their selection along three axes (X, Y and Z) displayed as the top box, middle box and bottom box under label Selected Analysis Dimensions in Figure 11.49. The available measures defined with the corresponding cube are also shown in the GUI, such as Average Rate, Internet Order Quantity and Internet Sales Amount.

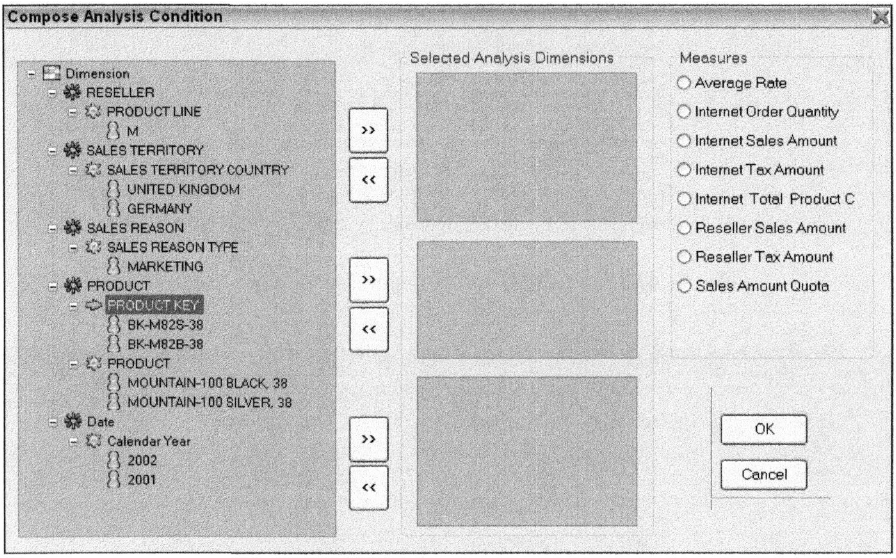

Fig. 11.49 MDX Analysis Condition Customization in Decision Cycle IV

A finished analysis condition customization is shown in Figure 11.50, which results in a MDX query as follows.

SELECT

{([*PRODUCT*].[*PRODUCT*].[*Mountain-100 Silver, 38*]),

([*PRODUCT*].[*PRODUCT*].[*Mountain-100 Black, 38*])

} ON 0,

{([*SALES TERRITORY*].[*SALES TERRITORY COUNTRY*].[*United Kingdom*]),

([*SALES TERRITORY*].[*SALES TERRITORY COUNTRY*].[Germany])

} ON 1,

{([*Date*].[*Calendar Year*].[*2001*]), ([*Date*].[*Calendar Year*].[*2002*])

} ON 2

FROM [*Adventure Works*]

WHERE [*Internet Sales Amount*]

This MDX query has there axes: 0, 1 and 2.

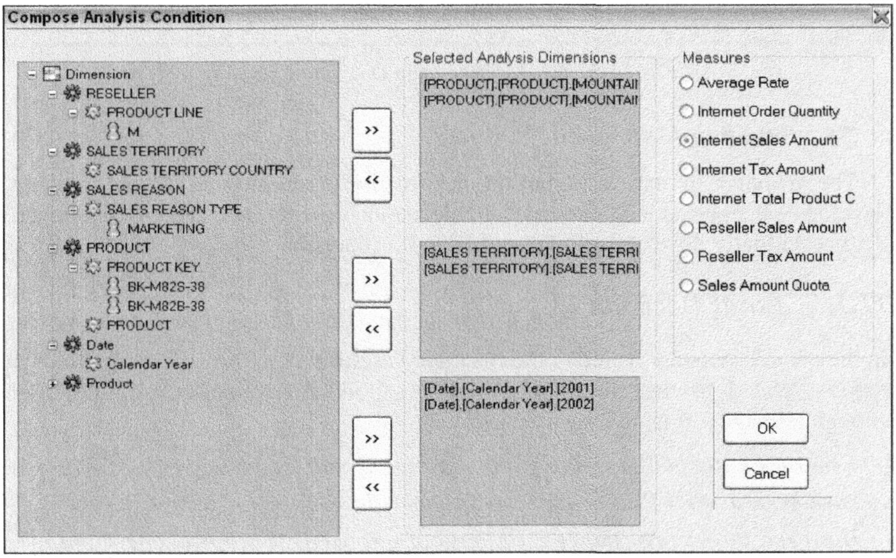

Fig. 11.50 Finished MDX Analysis Condition Customization in Decision Cycle IV

When the manager finishes customizing the analysis conditions, the corresponding MDX query will be generated and submitted to the data warehouse to retrieve relevant situation information. The retrieved situation information is presented to the manager in the form of MDX report. Figure 11.51 shows a MDX report generated by the MDX reporting module according to the analysis condition defined in Figure 11.50. This MDX report shows the Internet sales amount across the United Kingdom and Germany for different product models in the year 2002. There are two pages of reports generated for 2002 and 2001 respectively. Managers can click on the *Next Report* button to see different pages.

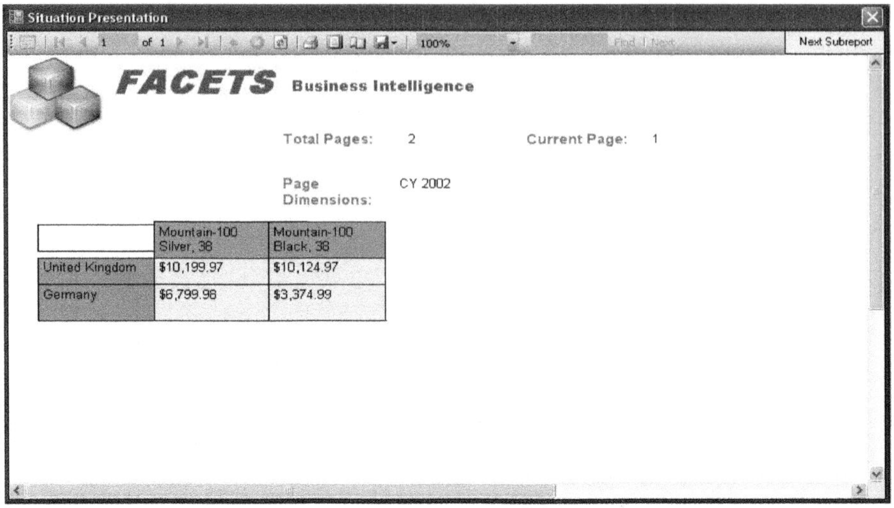

Fig. 11.51 A MDX Report in Decision Cycle IV

(7) Situation Assessment

The manager perceives situation information presented in the reports and develops a deeper understanding of the decision situation. As such, the manager's SA is eventually developed and enriched in this decision cycle.

11.1.6 Final Decision

In the end of decision cycle IV, the manager carefully read through all the relevant reports related to the current decision situation and found himself confident enough to make a final decision as follows.

Federal Express (FE) is one of the most important business partners. We have been using FE to deliver our products in Europe for over 10 years. However, things just changed. FE has outsourced his European operation to some local logistics companies due to the petrol price rise. The outsourced operation keeps the cost down, however product delivery has become very inefficient. Consequently, the average delivery time of BK-M82S-42 has risen from 3 business days to over 1 week. The over slow delivery service has discouraged our old customers and potential buyers. Therefore, urgently we need create new partnerships for product delivery in order to increase the sales of BK-M82S-42.

11.2 Application Case II: Public Health

In this section, we will briefly report a research project, conducted by a Master student (Alam 2009), in which FACETS was integrated into an early warning

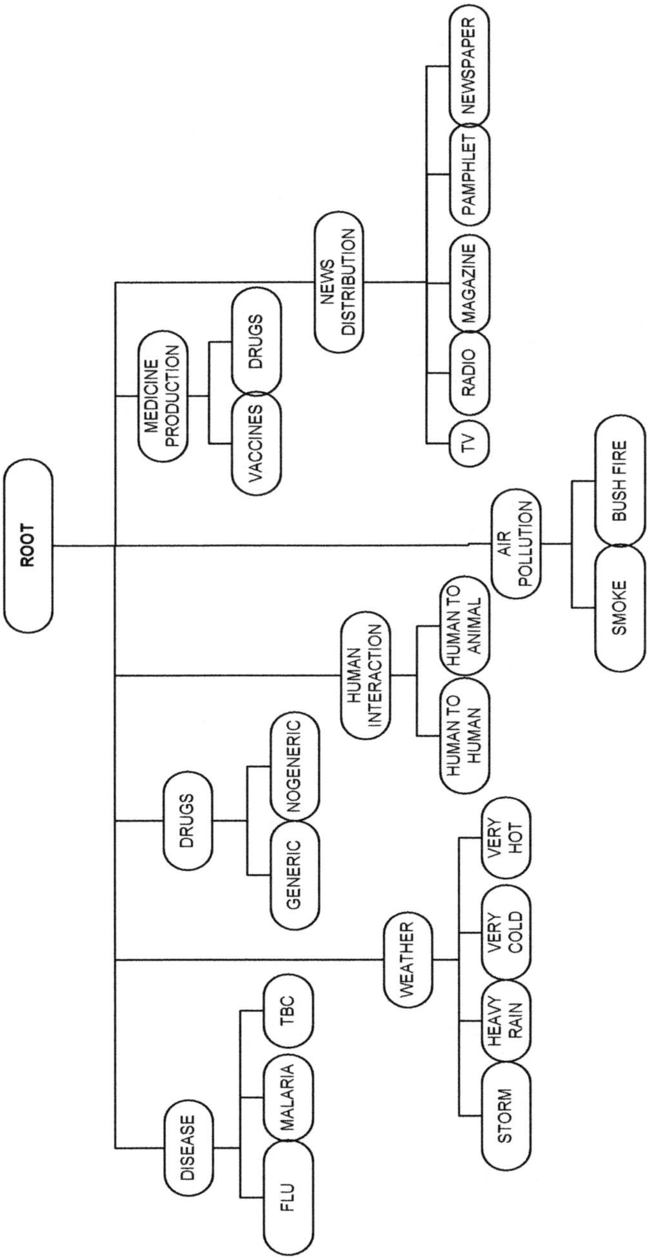

Fig. 11.52 The Class Tree of the EWS Ontology

system (EWS) in public health. The early warning system is used to support decision making for prediction, prevention and response to future pandemics.

Early warning systems are considered as effective tools to give prediction before the real problems occur. The EWS in pandemic control are complex and difficult to develop, due to some real issues, such as detecting the future pandemic, controlling all disease, late disease prevention, measuring all levels of disease, monitoring drug's stock and vaccine's stock, and predicting disease behaviors. In this project, FACETS was incorporated into an EWS as its key part for decision support.

According to the technical specification of FACETS, an ontology called *EWSO* was developed in order to apply FACETS in the context of pandemic early warning. In *EWSO*, 64 classes were defined, corresponding to 63 subsumption relationships in the class tree. Among these classes, 51 classes are correlated to tables or cubes in the data warehouse. Over 300 property-share relationships across these classes were also defined. Part of the class tree of *EWSO* is shown in Figure 11.52.

The experience base was created using the method of experience elicitation discussed in Section 6.2.2. Part of the experience base is shown in Figure 11.53, which represents how FLU is affected by different factors from different factors.

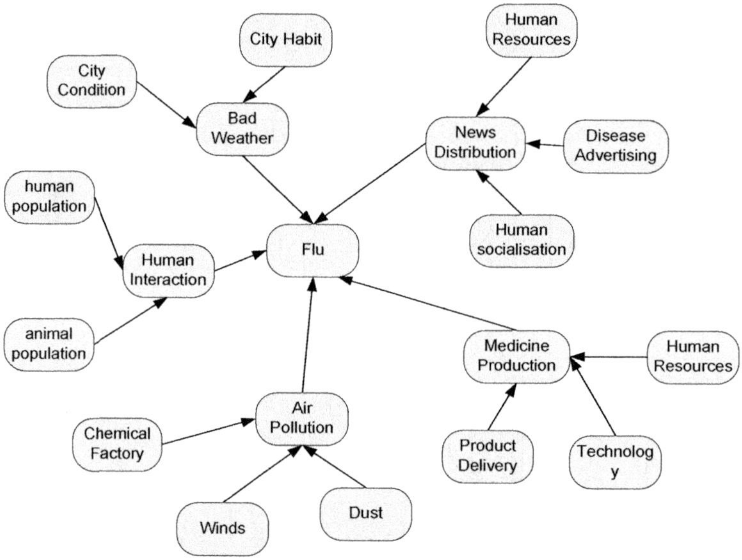

Fig. 11.53 The Mental Models of the EWS

This experience map will help a decision maker to approach the right disease information regarding his/her current decision situation. For example, for the issue of NEWS DISTRIBUTION, the relevant information that should be examined includes HUMAN RESOURCES, DISEASE ADVERTISING, and HUMAN SOCIALISATION. FACETS will utilize these causal relationships between different issues to assist the decision makers to expose valuable information for decision making.

A decision making problem supported by the EWS is described as follow:

How to catch up Flu Pandemic before it happen in Sydney?

Some situation descriptions of the decision maker when using FACETS in the decision process are as follows:

Influenza epidemics of type C in occurred twice in Broken Valley in 2001 and 2003.

We always see strong storms in the metropolitan areas of Broken Valley.

A newly constructed chemistry plant is causing serve air pollution in that area.

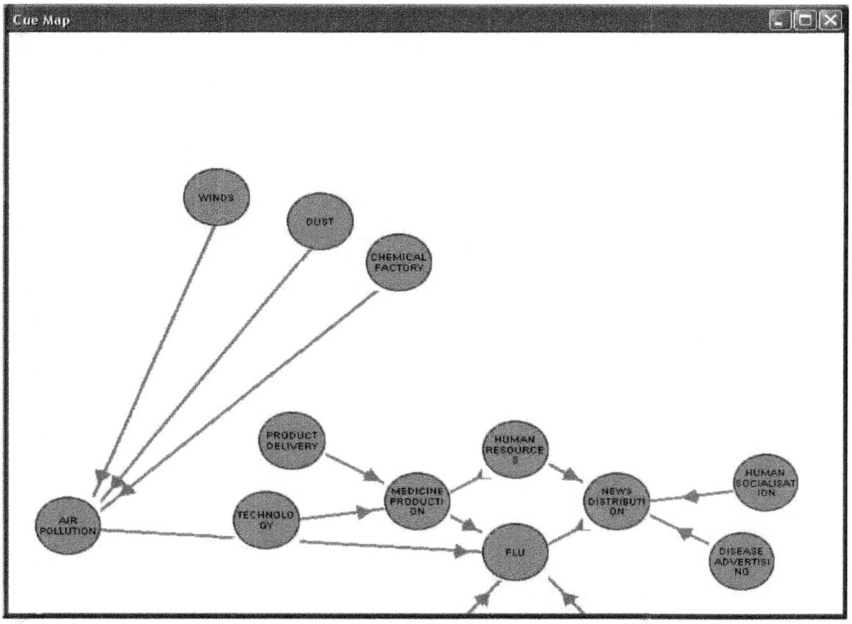

Fig. 11.54 A Cuemap Generated by FACETS in the EWS

The decision maker describes the decision situation in form of natural language, and inputs these descriptions into FACETS. The descriptions are analyzed according to the domain ontology and experiences. Corresponding data warehouse queries are constructed to retrieve situation information and present it to the decision maker according to the cue map. One of the cue maps generated by FACETS is shown in Figure 11.54. The decision maker browses the cue map and examines the situation information connected to different concepts. By this way, the decision maker is able to develop his/her SA about the current decision situation. The final solution in this application case is as follows.

(1) We need to plant more trees in Sydney.
(2) We need to give vaccines to all people.
(3) We need to spread the appropriate information to protect them from Flu.
(4) We need to manage the operation standard for all chemical factories.
(5) We need to manage all medicines stock.

11.3 Summary

FACETS is a decision support system, particularly for dealing with ill-structured problems in complex decision situations. This chapter presents two application cases of FACETS in two different domains: business and public health. No matter what decision situation, FACETS is an effective tool which assists the decision maker to approach the most relevant information to the decision situation. The basic methodology of decision support in FACETS is to seek and present useful situation information for the decision maker, and help the decision maker to develop rich situation awareness about the decision situation. Better situation awareness is more likely lead to better decisions, which is theoretical basis of FACETS.

As can be seen from the two application cases, the application of FACETS requires three components: a domain ontology, an experience base and a data warehouse, which are customized according to the application domain. Technically, FACETS can be used for decision support in any domain, as long as the appropriate domain ontology, experience base and data warehouse can be constructed.

References

Aamodt, A., Plaza, E.: Case-based reasoning: foundational issues, methodological variations, and system approaches. AI Communications 7(1), 39–59 (1994)

Ackermann, F., Eden, C., Cropper, S.: Getting started with cognitive mapping. In: The 7th Young OR Conference, University of Warwick. Coventry CV4 7AL, UK (1992)

Adler, M., Ziglio, E.: Gazing into the Oracle: The Delphi Method and Its Application to Social Policy and Public Health. Jessica Kingsley Publishers, London (1996)

Agrell, P.J.: A multicriteria framework for inventory control. International Journal of Production Economics 41, 59–70 (1995)

Alam, D.M.: Implementation of the ontology and mental model for hospital early warning systems (HEWS). University of Technology, Sydney (2009)

Alavi, M., Keen, P.G.W.: Business teams in an information age. The Information Society 64, 179–195 (1989)

Alter, S.: Decision Support Systems: Current Practice and Continuing Challenges. Addison-Wesley Pub., Reading (1980)

Amgoud, L., Belabbes, S., Prade, H.: Towards a formal framework for the search of a consensus between autonomous agents. In: Parsons, S., Maudet, N. (eds.) Argumentation in Multi-Agent Systems, pp. 264–278. Springer, Rahwan (2006)

Amgoud, L., Bonnefon, J.-F., Prade, H.: An argumentation-based approach to multiple criteria decision. In: Godo, L. (ed.) ECSQARU 2005. LNCS, vol. 3571, pp. 269–280. Springer, Heidelberg (2005)

Amgoud, L., Prade, H.: Handling threats, rewards and explanatory arguments in a unified setting. International Journal of Intelligent Systems 20, 1195–1218 (2005)

Amgoud, L., Prade, H.: Explaining qualitative decision under uncertainty by argumentation. In: Proc. Twenty-First National Conference on Artificial Intelligence (AAAI 2006), Boston, pp. 219–224 (2006)

Ana-Maria, P., Oren, E., Henry, K.: Towards a theory of natural language interfaces to databases. In: Proceedings of the 8th international conference on Intelligent user interfaces. ACM, Miami (2003)

Anandalingam, G., Friesz, T.: Hierarchical optimization: An introduction. Annals of Operations Research 34, 1–11 (1992)

Androutsopoulos, I., Ritchie, G., Thanisch, P.: Natural language interfaces to databases–an introduction. Journal of Language Engineering 1(1), 29–81 (1995)

Angehrn, A.A., Jelassi, T.: DSS research and practice in perspective. Decision Support Systems 12, 267–275 (1994)

Anthony, R.N.: Planning and control systems: a framework for analysis. Harvard University, Cambridge (1965)

Antunes, C.H., Almeida, L.A., Lopes, V., Climaco, J.N.: A decision support system dedicated to discrete multiple criteria problems. Decision Support Systems 12, 327–335 (1994)

Arnott, D., Pervan, G.: A critical analysis of decision support systems research. Journal of Information Technology 20(2), 67–87 (2005)

Ba, S., Kalakota, R., Whinston, A.B.: Using client-broker-server architecture for intranet decision support. Decision Support Systems 19, 171–192 (1997)

Bahrick, H.P., Bahrick, P.E., Wittlinger, R.P.: Fifty years of names and faces: A cross sectional approach. Journal of Experimental Psychology: General 104, 54–75 (1975)

Ballou, R.: Business Logistics/Supply Chain Management. Pearson Prentice Hall, London (2004)

Barclay, S.: A User's Manual to HIVIEW. London School of Economics, London (1987)

Barclay, S.: A User's Manual to EQUITY. London School of Economics, London (1988)

Barr, S., Sharda, R.: Effectiveness of decision support systems - development or reliance effect. Decision Support Systems 21, 133–146 (1997)

Baum, E.: What is thought. MIT Press, Cambridge (2004)

Bazerman, M.H.: Judgment in Managerial Decision Making, 6th edn. Wiley, New York (2005)

Belkin, N.J., Oddy, R.N., Brooks, H.M.: Ask for information-retrieval. 1. Background and theory. Journal of Documentation 38(2), 61–71 (1982)

Bellinger, G., Castro, D., Mills, A.: Data, Information, Knowledge, and Wisdom (2004), `http://www.systems-thinking.org/dikw/dikw.htm` (retrieved July 14, 2007)

Bellman, R.E., Zadeh, L.A.: Decision-making in a fuzzy Environment. Management Science 17, B141–B164 (1970)

Belton, V., Vickers, S.P.: Visual Interactive Sensitivity Analyses for Multiple Criteria Decision Analysis: User Manual, SPV Software Products, UK (1989)

Benayoun, R., Montogolfier, J., Tergny, J., Larichev, O.: Linear programming with multiple objective functions: step method (STEM). Mathematical Programming 1, 366–375 (1971)

Bergmann, R.: Experience management: Foundations, development methodology, and internet-based applications. Springer, Berlin (2002)

Berry, M.J.A., Linoff, G.: Data Mining Techniques. John Wiley & Sons, New York (1997)

Bhargava, H., Krishnan, R., Muller, R.: Decision support on demanded: emerging electronic markets for decision technologies. Decision Support Systems 19, 193–214 (1997)

Bidgoli, H.: Decision Support Systems: Principles and Practice. West Pub. Co., St. Paul (1989)

Bois, P.D., Brans, J.P., Cantraine, F., Mareschal, B.: Medics: an expert system for computer-aided diagnosis using the PROMETHEE multicriteria method. European Journal of Operational Research 39, 284–292 (1989)

Bonczek, R.H., Holsapple, C.W., Whinston, A.B.: Foundations of Decision Support Systems. Academic Press, New York (1981)

Borenstein, D.: Idssflex - an intelligent DSS for the design and evaluation of flexible manufacturing systems. Journal of the Operational Research Society 49, 734–744 (1998)

Bose, U., Davey, A.M., Olson, D.L.: Multi-attribute utility methods in group decision making: past applications and potential for inclusion in GDSS. Omega 25, 691–706 (1997)

Bouchon-Meunier, B., Yager, R.R., Zadeh, L.A.: Information, Uncertainty, Fusion. Kluwer Academic Publishers, Norwell (2000a)

Bouchon-Meunier, B., Yager, R.R., Zadeh, L.A.: Uncertainty in Intelligent and Information Systems. World Scientific Publishers, Singapore (2000b)

Boufaden, N.: An ontology-based semantic tagger for IE system. In: The 41st Annual Meeting on Association for Computational Linguistics. Association for Computational Linguistics, Sapporo (2003)

Brito, M.P.D., Dekker, R.: Reverse Logistics - A Framework, Econometric Institute Report 290, Erasmus University Rotterdam, Econometric Institute (2002)

Bubnicki, Z.: Uncertain Logics, Variables and Systems. Springer, Berlin (2002)

Budgen, D., Marashi, M.: MDSE advisor: knowledge-based techniques applied to software design assessment. Knowledge-Based Systems 1, 235–239 (1988)

Bui, T.: Co-oP: A Group Decision Support System for Cooperative Multiple Criteria Group Decision-Making. Springer, Berlin (1989)

Bui, T.X., Jelassi, T.M., Shakun, M.F.: Group decision and negotiation support systems (GDNSS). European Journal of Operational Research 46, 141–142 (1990)

BusinessObjects, BusinessObjects Polestar (2008), `http://www.businessobjects.com/product/catalog/polestar/` (retrieved August 15, 2008)

Calpine, H.C., Golding, A.: Some properties of Pareto-optimal choices in decision problems. Omega 4, 141–147 (1976)

Carley, K., Palmquist, M.: Extracting, representing and analyzing mental models. Social Forces 70(3), 601–636 (1992)

Carlson, D.A., Ram, S.: HyperIntelligence: the next frontier. Communications of the ACM 33(3), 311–321 (1990)

Carlsson, C., Fuller, R.: Adaptive fuzzy cognitive maps for hyperknowledge representation in strategy formation process. Paper presented at the International Panel Conference on Soft and Intelligent Computing, Budapest, Hungary (1996)

Caserta, J.: Practical techniques for extracting, cleaning, conforming, and delivering data. John Wiley & Sons, Chichester (2004)

Castells, P., Fernández, M., Vallet, D.: An adaptation of the vector-space model for ontology-based information retrieval. IEEE Transactions on Knowledge and Data Engineering 17(2), 261–272 (2007)

Chan, F.T.S., Chan, M.H., Tang, N.K.H.: Evaluation methodologies for technology selection. Journal of Materials Processing Technology 107, 330–337 (2000)

Chankong, V., Haimes, Y.Y.: Multiobjective Decision Making: Theory and Methodology. North-Holland, Amsterdam (1983)

Charnes, A., Cooper, W.W.: Management Models and Industrial Applications of Linear Programming, vol. I and II. Wiley, Chichester (1961)

Charnes, A., Cooper, W.W.: Goal programming and multiple objective optimizations. European Journal of Operational Research 1, 39–54 (1977)

Chaudhry, S.S., Salchenberger, L., Beheshtian, M.: A small business inventory DSS - design, development, and implementation issues. Computers & Operations Research 23, 63–72 (1996)

Chen, J.Q., Lee, S.M.: An exploratory cognitive DSS for strategic decision making. Decision support systems 36, 147–160 (2003)

Chen, T.: Decision Analysis. Science Press, Beijing (1987)
Chen, Z.X.: Computational Intelligence Decision Support System. University of Nebraska, Omaha (2000)
Cheng, C.-H., Yang, K.-L., Hwang, C.-L.: Evaluating attack helicopters by AHP based on linguistic variable weight. European Journal of Operational Research 116, 423–435 (1999)
Chidambaram, L.: Relational development in computer-supported groups. MIS Quarterly 20, 143–163 (1996)
Chuang, T.T., Yadav, S.B.: The development of an adaptive decision support system. Decision Support Systems 24, 73–87 (1998)
Clarke, S.J., Willett, P.: Estimating the recall performance of Web search engines. Aslib Proceedings 49(7), 184–189 (1997)
Codd, E.F., Codd, S.B., Salley, C.T.: Providing OLAP (On-line Analytical Processing) to user-analysts: An IT mandate (1993), `http://www.fpm.com/refer/codd.html` (retrieved June 6, 2008)
Cohon, J.L.: Multiobjective Programming and Planning. Academic Press, New York (1978)
Colson, G., Mareschal, B.: JUDGES: a descriptive group decision support system for the ranking of items. Decision Support Systems 12, 391–404 (1994)
Connell, N.A.D., Powell, P.L.: A comparison of potential applications of expert systems and decision support systems. Journal of the Operational Research Society 41, 431–439 (1990)
Cook, W.D., Kress, M.: Ordinal ranking with intensity of preference. Management science 31, 26–32 (1985)
Courtney, J.F.: Decision making and knowledge management in inquiring organizations: toward a new decision-making paradigm for DSS. Decision Support Systems 31, 17–38 (2001)
Csaki, P., Rapcsak, T., Turchanyi, P., Vermes, M.: R and D for group decision aid in Hungary by WINGDSS, a Microsoft Windows based group decision support system. Decision Support Systems 14, 205–217 (1995)
CSIRO. Business intelligence - What is it? (2003, May 28, 2008), `http://www.cmis.csiro.au/bi/what-is-BI.htm` (retrieved April 8, 2009)
Daily, B., Whatley, A., Ash, S.R., Steiner, R.L.: The effects of a group decision support system on culturally diverse and culturally homogeneous group decision making. Information and Management 30, 281–289 (1996)
Damousis, I.G., Bakirtzis, A.G., Dokopoulos, P.S.: Network-constrained economic dispatch using real-coded genetic algorithm. IEEE Transactions on Power Systems 18, 198–205 (2003)
Davey, A., Olson, D.: Multiple criteria decision making models in group decision support. Group Decision and Negotiation 7, 55–75 (1998)
Dawes, R.M.: Social selection based on multidimentional criteria. Journal of Abnormal and Social Psychology 68, 104–109 (1964)
Dawes, R.M.: Rational choice in a uncertain world. Harcourt Brace Jovanoich, New York (1988)
Degoulet, P., Fieschi, M., Chatellier, G.: Decision support systems from the standpoint of knowledge representation. Methods of Information in Medicine 34, 202–208 (1995)
Dennis, A.R.: Information exchange and use in group decision making: you can lead a group to information, but you can't make it think. MIS Quarterly 20, 433–457 (1996)

Department of Communicable Disease (Who), WHO guidelines for the global surveillance of severe acute respiratory syndrome, SARS (2004), http://www.who.int/csr/resources/publications/WHO_CDS_CSR_ARO_2004_1/en/index.html

Despres, S., Rosenthal, S.C.: Designing decision support systems and expert systems with a better end-use involvement: a promising approach. European Journal of Operational Research 61, 145–153 (1992)

Ditsa, G.: Executive information systems use in organisational contexts: an explanatory user behaviour testing. In: Ditsa, G. (ed.) Information management: support systems & multimedia technology, pp. 109–155. USA Idea Group Publishing, Hershey (2003)

Donaldson, G., Lorsch, J.W.: Decision making at the top: The shaping of strategic direction. Basic Books, New York (1983)

Downing, C.E., Ringuest, J.L.: Implementing and testing a complex interactive MOLP algorithm. Decision Support Systems 33, 363–374 (2002)

Dubois, D., Fargier, H.: Qualitative decision rules under uncertainty. In: Della Riccia, G., Dubois, D., Kruse, R., Lenz, H.-J. (eds.) Planning Based on Decision Theory, pp. 3–26. Springer, Wien (2004)

Dubois, D., Fargier, H., Perny, P.: Qualitative decision theory with preference relations and comparative uncertainty: an axiomatic approach. Artificial Intelligence 148, 219–260 (2003)

Dubois, D., Fargier, H., Perny, P., Prade, H.: A characterization of generalized concordance rules in multicriteria decision making. International Journal of Intelligent Systems 18, 751–774 (2003)

Dubois, D., Fargier, H., Prade, H., Perny, P.: Qualitative decision theory: from savage's axioms to nonmonotonic reasoning. Journal of the ACM 49, 455–495 (2002)

Dubois, D., Fortemps, P.: Computing improved optimal solutions to max-min flexible constraint satisfaction problems. European Journal of Operational Research 118, 95–126 (1999)

Dubois, D., Fortemps, P.: Selecting preferred solutions in the minimax approach to dynamic programming problems under flexible constraints. European Journal of Operational Research 160, 582–598 (2005)

Dubois, D., Hüllermeier, E., Prade, H.: Fuzzy methods for case-based recommendation and decision support. Journal of Intelligent Information Systems 27, 95–115 (2006)

Dunham, M.: Data Mining. Prentice Hall, Upper Saddle River (2003)

Dyer, J.S.: Interactive goal programming. Management Science 19, 62–70 (1972)

EconomistIntelligenceUnit, What do companies want from business intelligence? (Vol 2006): Economist Intelligence Unit (2006)

Eden, C.: Cognitive mapping. European Journal of Operational Research 36(1), 1–13 (1988)

Eden, C., Ackermann, F.: Horses for courses - a stakeholder approach to the evaluation of GDSS. Group Decision and Negotiation 5, 501–519 (1996)

Edwards, J.S.: Expert systems in management and administration - are they really different from decision support systems? European Journal of Operational Research 61, 114–121 (1992)

Emerson, T.J., Reising, J.M., Britten-Austin, H.G.: Workload and situation awareness in future aircraft. SAE Technical Paper, No. 871803 (1987)

Endsley, M.R.: Measurement of situation awareness in dynamic systems. Human Factors 37(1), 65–84 (1995a)

Endsley, M.R.: Toward a theory of situation awareness in dynamic systems. Human Factors 37, 32–64 (1995b)

Endsley, M.R.: Theoretical underpinnings of situation awareness: a critical review. In: Endsley, M.R., Garland, D.J. (eds.) Situation Awareness Analysis and Measurement. Lawrence Erlbaum Associates, Mahwah (2000)

Endsley, M.R., Garland, D.: Situation Awareness Analysis and Measurement. Lawrence Erlbaum Associates, Publishers, Mahway (2000)

Endsley, M.R., Bolte, B., Jones, D.G.: Designing for situation awareness: an approach to user-centered design, 1st edn. Taylor&Francis, London (2003)

Engemann, K.J., Holmes, M.E., Yager, R.R.: Decision making with attitudinal expected values. International Journal of Technology, Policy and Managemant 4, 353–364 (2004)

Eom, S.B.: Relationships between the decision support system subspecialties and reference disciplines: an empirical investigation. European Journal of Operational Research 104, 31–45 (1998)

Er, M.C., Ng, A.C.: The anonymity and proximity factors in group decision support systems. Decision Support Systems 14, 75–83 (1995)

Ertay, T., Ruan, D.: Data envelopment analysis based decision model for optimal operator allocation in CMS. European Journal of Operational Research 164, 800–810 (2005)

Ertay, T., Ruan, D., Tuzkaya, U.R.: Integrating data envelopment analysis and analytic hierarchy for the facility layout design in manufacturing systems. Information Sciences 176, 237–262 (2006)

Evelson, B., Moore, C., Kobielus, J., Karel, R., Nicolson, N.: The Forrester wave: Enterprise business intelligence platforms, Q3 (2008)

Expert-Choice-Inc, Expert Choice: Walk-through, Pittsburgh, USA (1992)

Farmer, T.A.: Testing the robustness of multiattribute utility theory in an applied setting. Decision Sciences 18, 178–193 (1987)

Fetzer, J.H.: People are not computers: (most) though processes are not computational procedures. Journal of Experimental and Theoretical Artificial Intelligence 10, 371–391 (1998)

Finlay, P.N.: Decision support systems and expert systems: a comparison of their components and design methodologies. Computers and Operations Research 17, 535–543 (1990)

Giaglis, G.M., Paul, R.J., Doukidis, G.I.: Dynamic modelling to assess the business value of electronic commerce. International Journal of Electronic Commerce 3, 35–51 (1999)

Giordani, A.: Mapping natural language into SQL in a NLIDB. In: Kapetanios, E., Sugumaran, V., Spiliopoulou, M. (eds.) Natural Language and Information Systems, pp. 367–371. Springer, Heidelberg (2008)

Giuntini, R., Andel, T.: Advance with reverse logistics. Transportation and Distribution, Part 1 36, 93–98 (1995a)

Giuntini, R., Andel, T.: Reverse logistics role models. Transportation and Distribution, Part 3 36, 97–98 (1995b)

Gnyawali, D.R., Tyler, B.B.: Cause mapping in strategic management research: processes, issues, and observations. In: David, J., Ketchen, J., Bergh, D.D. (eds.) Research methodology in strategic and management, vol. 2, pp. 225–257. Elsevier, San Diego (2005)

Gordon, L.A., Pinches, G.E.: Improving capital budgeting: a decision support system approach (1984)

Gorry, G.A., Morton, M.S.S.: A framework for management information systems. Sloan Management Review 13(1), 50–70 (1971)

Gomez-Perez, A., Fernandez-Lopez, M., Corcho, O.: Ontological Engineering. Springer, Heidelberg (2004)

Gooley, T.B.: Reverse logistics: five steps to success. Logistics Management and Distribution Report 37, 49–55 (1998)

Gorry, G.A., Scott-Morton, M.S.: A framework for management information system. Sloan Management Review 13, 55–70 (1971)

Gottinger, H.W., Weimann, P.: Intelligent decision support systems. Decision Support Systems 8, 317–332 (1992)

Gray, P.: Group decision support systems. Decision Support Systems 3, 233–242 (1987)

Gray, P., Mandviwalla, M.: New direction for GDSS. Group Decision and Negotiation 8, 77–83 (1999)

Gruman, G.: Rethinking business intelligence. InfoWorld 29(14), 22–27 (2007)

Guarino, N.: Formal ontology in Information Systems. Paper presented at the the 1st International Conference on Formal Ontologies in Information Systems, Trento, Italy (1998)

Guerlian, S., Brown, C.E., Mastrangelo, C.: Intelligent decision support systems. In: Proc. IEEE International Conference on Systems, Man, and Cybernetics, pp. 1934–1938 (2000)

Guyonnet, D., Bourgine, B., Dubois, D., Fargier, H., Côme, B., Chilès, J.-P.: Hybrid approach for addressing uncertainty in risk assessments. Journal of Environmental Engineering 129, 68–78 (2003)

Hamilton, W.L.: Situation awareness metrics program. SAE Technical Paper(871767) (1987)

Han, J., Kamber, M.: Data Mining: Concepts and Techniques. Morgan Kaufmann, San Francisco (2001)

Haseman, W.D., Nazareth, D.L., Paul, S.: Implementation of a group decision support system utilizing collective memory. Information & Management 42, 591–605 (2005)

Herrera, F., Herrera-Viedma, E.: Choice functions and mechanisms for linguistic preference relations. European Journal of Operational Research 120, 144–161 (2000)

Hipel, K.: Multiple objective decision making in water resources. Water Resources Bulletin 28, 3–12 (1992)

Hoffman, J.: Introduction to Structured Query Language (2001, 4-9-2001), `http://riki-lb1.vet.ohio-state.edu/mqlin/computec/tutorials/SQLTutorial.htm` (retrieved September 11, 2008)

Holsapple, C.W., Whinston, A.B.: Decision Support Systems: A Knowledge-Based Approach. West Pub. Co., Minneapolis, St. Paul (1996)

Hoven, J.V.D.: Executive support systems & decision making. Journal of Systems Management 47(2), 48–55 (1996)

Hu, T.L., Sheu, J.B., Huang, K.H.: A reverse logistics cost minimization model for the treatment of hazardous wastes. Transportation Research, Part E 38, 457–473 (2002)

Hwang, C.L., Lai, Y.J., Liu, D.R.: A new approach for multiple objective decision making. Computer & Operations Research 20, 889–899 (1993)

Hwang, C.L., Masud, A.S.: Multiple Objective Decision Making: Methods and Applications. Springer, Berlin (1979)

Hwang, C.L., Yoon, K.: Multiple Attribute Decision Making: Methods and Applications - A State of the Art Survey. Springer, Berlin (1981)

Ignizio, J.P.: Goal Programming and Extensions. Lexington Books, Massachusetts (1976)

Ignizio, J.P.: The determination of a subset of efficient solutions via goal programming. Computer and Operations Research 8, 9–16 (1981)

Ignizio, J.P.: Generalized goal programming: an overview. Computer & Operations Research 10, 277–289 (1983)

Ingwersen, P.: Information retrieval interaction. Tayor Graham, London (1996)
Inmon, W.H.: What is a data warehouse. PRISM Newsletter 1(1) (1993)
Inmon, W.H.: Building the data warehouse, 3rd edn. Wiley, New York (2002)
Isenberg, D.J.: How senior manager think. Harvard Business Review, 82–90 (November – December 1984)
Iz, P.H.: Two multiple criteria group decision support systems based on mathematical programming and ranking methods. European Journal of Operational Research 61, 245–253 (1992)
Iz, P.H., Gardiner, L.R.: Analysis of multiple criteria decision support systems for cooperative groups. Group Decision and Negotiation 2, 61–79 (1993)
Jacob, V.S., Pirkul, H.: Organizational decision support systems. International Journal of Man-Machine Studies 36, 817–832 (1992)
Barnes, J.H.: Cognitive biases and their impact on strategic planning. Strategic Management Journal 5(2), 129–137 (1984)
Jaques, E.: A general theory of bureaucracy. Gower Publishing, Hampshire (1976)
Jaques, E.: Requisite organization. Cason Hall, Virginia (1996)
Jayabarath, T., Jayaprakash, K., Jeyakumar, D.N., Raghunathan, T.: Evolutionary programming techniques for different kinds of economic dispatch problems. Electric Power Systems Research 73, 169–176 (2005)
Jelassi, M.T., Beauclair, R.A.: An integrated framework for group decision support design. Information & Management 13, 143–153 (1987)
Jelassi, M.T., Jarke, M., Stohr, E.A.: Designing a generalized multiple criteria decision support system. Journal of Management Information Systems 1, 24–43 (1985)
Jensen, P.A., Bard, J.F.: Operations research: models and methods Hoboken. Wiley, New Jersey (2003)
Johnson-Laird, P.N., Girotto, V., Legrenzi, P.: Mental models: a gentle guide for outsiders. Sistemi Intelligenti 9, 68–33 (1998)
Juhani, I.: An empirical test of the DeLone-McLean model of information system success. SIGMIS Database 36, 8–27 (2005)
Jung, H., Lee, G.G.: Multilingual Question Answering with High Portability on Relational Databases. In: The 19th International Conference on Computational Linguistics, Taipei (2002)
Kacprzyk, J., Fedrizzi, M., Nurmi, H.: Group decision making and consensus under fuzzy preference and fuzzy majority. Fuzzy Sets and Systems 49, 21–31 (1992)
Kasper, G.M.: A theory of decision support system design for user calibration. Information Systems Research 7, 215–232 (1996)
Keen, P.G.W.: Interactive computer systems for managers: a modest proposal. Sloan Management Review 18, 1–17 (1976)
Keen, P.G.W., Scott Morton, M.S.: Decision Support Systems: An Organizational Perspective. Addison-Wesley Pub. Co., Reading (1978)
Kepner, C.H., Tregoe, B.B.: Problem Analysis and Decision Making. Kepner-Tregoe Ltd., Princeton (1973)
Kersten, G.E.: NEGO - Group decision support system. Information & Management 8, 237–246 (1985)
Klein, G.A., Calderwook, R., Macgregor, D.: Critical decision method for eliciting knowledge. IEEE Transactions on Systems, Man, and Cybernetics - Part A: System and Humans 19, 462–472 (1989)
Klein, M.R., Methlie, L.B.: Knowledge-Based Decision Support Systems with Applications in Business. John Wiley & Sons, Chichester (1995)

Klimberg, R.: GRADS: A new graphical display system for visualizing multiple criteria solutions. Computers and Operations Research 19, 707–711 (1992)

Knoblock, C.A., Lerman, K., Minton, S., Muslea, I.: Accurately and reliably extracting data from the web: A machine learning approach. IEEE Data Engineering Bulletin 23(4), 33–41 (2000)

Kolodneer, J.L.: Improving human decision making through case-based decision aiding. AI Magazine 12, 52–68 (1991)

Konar, A., Chakraborty, U.K.: Reasoning and unsupervised learning in a fuzzy cognitive map. Information Sciences 170, 419–441 (2005)

Korhonen, P., Lewandowski, A., Wallenius, J.: Multiple Criteria Decision Support. Lecture Notes in Economics and Mathematical Systems, vol. 356. Springer, Berlin (1991)

Korhonen, P., Wallenius, J.: A Multiple objective linear programming decision support system. Decision Support Systems 6, 243–251 (1990)

Kosko, B.: Fuzzy cognitive maps. International Journal of Man-Machine Studies 24, 65–75 (1986)

Kuo, F.-Y.: Managerial intuition and the development of executive support systems. Decision Support Systems 24, 89–103 (1998)

Lagreze, E.J., Shakun, M.F.: Decision support systems for semi-structured buying decisions. European Journal of Operational Research 16, 48–58 (1984)

Lahti, R.K.: Group decision making within the organization: can models help? CSWT paper, `http://www.eorkteams.unt.edu/reports/lahti.htm`

Lai, Y.J.: Imost - interactive multiple objective system technique. Journal of the Operational Research Society 46, 958–976 (1995)

Lai, Y.J., Liu, T.Y., Hwang, C.L.: Topsis for MODM. European Journal of Operational Research 76, 486–500 (1994)

Lambert, D.M., Cooper, M.C.: Issues in supply chain management. Industrial marketing Management 29, 65–83 (2000)

Langfield-Smith, K., Wirth, A.: Measuring differences between cognitive maps. The Journal of the Operational Research Society 43(12), 1135–1150 (1992)

Leacock, C., Chodorow, M.: Combining local context and WordNet similarity for word sense identification. In: Fellbaum, C. (ed.) Wordnet: an electronic lexical database, pp. 265–283. MIT Press, Cambridge (1998)

Lee, K.C., Kim, H.S.: A fuzzy cognitive map-based bi-directional inference mechanism: an application to stock investment analysis. International Journal of Intelligent Systems in Accounting, Finance & Management 6, 41–57 (1997)

Lee, M., Pham, H., Zhang, X.: A methodology for priority setting with application to software development process. European Journal of Operational Research 118, 375–389 (1999)

Lee, S.M.: Goal Programming for Decision Analysis. Auerbach Publishers, Philadelphia (1972)

Li, Y., Yang, H., Jagadish, H.V.: Constructing a generic natural language interface for an XML database. In: Ioannidis, Y., Scholl, M.H., Schmidt, J.W., Matthes, F., Hatzopoulos, M., Böhm, K., Kemper, A., Grust, T., Böhm, C. (eds.) EDBT 2006. LNCS, vol. 3896, pp. 737–754. Springer, Heidelberg (2006)

Liberatore, M., Nydick, R.: Decision Technology: Modeling, Software, and Applications. Wiley, New Jersey (2003)

Lin, T.S., Yao, Y.Y., Zadeh, L.A.: Data Mining, Rough Sets and Granular Computing. Physica-Verlag, Heidelberg (2002)

Lipshitz, R., Klein, G., Orasanu, J., Salas, E.: Taking stock of naturalistic decision making. Journal of Behavioral Decision Making 14, 331–352 (2001)

Liu, J., Ruan, D., Carchon, R.: Synthesis and evaluation analysis of the indicator information in nuclear safeguards applications by computing with words. Int. J. Appl. Math. Comput. Sci. 12, 449–462 (2002)

Lu, J.: A Framework and Prototype for Intelligent Multiple Objectives Group Decision Support Systems, PhD Thesis, Curtin University of Technology, Australia (2000)

Lu, J., Lu, Z.: Development, distribution and evaluation for online tourism services in China. Electronic Commerce Research Journal 4, 221–239 (2004)

Lu, J., Quaddus, M.A.: An intelligence-based multi-objective group decision support system and its application. In: Proc. Information Technology for Business Management, 16th World Computer Congress (WCC 2000), Beijing, China, pp. 744–751 (2000)

Lu, J., Quaddus, M.A.: Integrating a knowledge based guidance system with multiple objective decision making. The New Zealand Journal of Applied Computer and Information Technology (NZJACIT) 5, 53–59 (2001a)

Lu, J., Quaddus, M.A.: A prototype of multi-objective group decision support systems with a group aggregation method base. In: Proc. the Twelfth Australasian Conference on Information Systems (ACIS), Coffs Harbour, Australia, pp. 378–394 (2001b)

Lu, J., Quaddus, M.A.: Experimental evaluation of an intelligent multiple objective group decision support system (IMOGDSS). In: Proc. the 9th European conference on IT evaluation, Paris, France, pp. 273–279 (2002)

Lu, J., Ruan, D., Zhang, G.: E-Service Intelligence: Methodologies, Technologies and Applications. Springer, Heidelberg (2006)

Lu, J., Ruan, D., Zhang, G., Zimmermann, H.J.: Editorial Preface: Special Issue on E-Service Intelligence. International Journal of Intelligent Systems 22, 397–400 (2006)

Lu, J., Tang, S., Mccullough, G.: An assessment for internet-based electronic commerce development in businesses of New Zealand. Electronic Markets: International Journal of Electronic Commerce and Business Media 11, 107–115 (2001)

Lu, J., Zhang, G.: Cost benefit factor analysis in e-services. International Journal of Service Industry Management (IJSIM) 14, 570–595 (2003a)

Lu, J., Zhang, G.: A model for evaluating e-commerce based on cost/benefit and customer satisfaction. Journal of Information Systems Frontiers 5, 265–277 (2003b)

Lu, J., Zhang, G., Ruan, D., Wu, F.: Multi-objective group decision making: Methods, software and applications with fuzzy set technology. Imperial College Press, London (2007)

Lu, J., Zhang, G., Shi, C.: Framework and implementation of a web-based WMODSS. In: Proc. The Workshop on Applications, Products and Services of Web-based Support Systems, in conjunction with the 2003 IEEE/WIC International Conference on Web Intelligence, Halifax, Canada, pp. 7–11 (2003)

Lu, J., Zhang, G., Wu, F.: Web-based multi-criteria group decision support system with linguistic term processing function. The IEEE Intelligent Informatics Bulletin 5, 35–43 (2005)

Lycan, W.G., Prinz, J.J.: Mind and cognition: An anthology, 3rd edn. Blackwell Publishers, Inc., Malden (2008)

Lyles, M.A., Thomas, H.: Strategic problem formulation: biases and assumptions embedded in alternative decision-making models. Journal of Management Studies 25, 131–145 (1988)

MacFadyen, H.: The parts of speech (2007, 2007.08.16), http://www.uottawa.ca/academic/arts/writcent/hypergrammar/partsp.html (retrieved January 12, 2008)
Mallach, E.G.: Decision Support and Data Warehouse Systems. McGraw Hill Higher Education, New York (2000)
Management, A.F.N.: Gathering Perceptions about the Organization Evaluation of Programs, http://www.allianceonline.org/FAQ/strategic_planning/what_is_situation_assessment.faq
March, J.G., Simon, H.: Organizations, 4th edn. Wiley, New York (1963)
Matheus, C.J., Kokar, M.M., Baclawski, K.: A core ontology for situation awareness. Paper presented at The 6th International Conference on Information Fusion, Cairns, Queensland, Australia (2003)
Maynard, D., Yankova, M., Kourakis, A., Kokossis, A.: Ontology-based information extraction for market monitoring and technology watch. Paper presented at the ESWC Workshop "End User Apects of the Semantic Web", Heraklion, Crete (2005)
McBride, N.: The rise and fall of an executive information system: a case study. Information Systems Journal 7, 4 (1997)
Mccarthy, J.: Applications of circumscription to formalizing common sense knowledge. Artificial Intelligence 28, 89–116 (1986)
McGuinness, D.: Ontologies and online commerce. IEEE Intelligent Systems 16, 8–14 (2001)
Microsoft, Adventure Works Cycles business scenarios (2007a), http://msdn.microsoft.com/en-us/library/ms124825.aspx (retrieved September 6, 2007)
Microsoft, Multidimensional Expressions (MDX) reference (2007b), http://msdn.microsoft.com/en-us/library/ms145506SQL.90.aspx (retrieved September 11, 2008)
Microsoft, Transact-SQL reference (2007c), http://msdn.microsoft.com/en-us/library/ms189826SQL.90.aspx (retrieved September 11, 2008)
Minch, R.P., Sanders, G.L.: Computerized information systems supporting multicriteria decision making. Decision Sciences 17, 395–413 (1986)
Mintzberg: The nature of managerial work. Harper & Row, New York (1973)
Mintzberg, H., Raisinghani, D., Théorêt, A.: The structure of "unstructured" decision processes. Administrative Science Quarterly 21, 246–275 (1976)
Missikoff, M., Schiappelli, F.: A method for ontology modeling in the business domain. Enterprise Modelling and Ontologies for Interoperability, Porto, Portugal (2005)
Mitra, S., Acharya, T.: Data Mining: Multimedia, Soft Computing and Bioinformatics. Wiley, New York (2003)
Moore, C.M.: Group Techniques for Ideal Building. Sage Publications, Newbury Park (1987)
Moreau, E.M.F.: The impact of intelligent decision support systems on intellectual task success: an empirical investigation. Decision Support Systems 42, 593–607 (2006)
Morehouse, B. Group systems corporation: corporate overview, http://www.groupsystems.com
Nazareth, D.L.: Integrating MCDM and DSS: barriers and counter strategies. Infor. 31, 1–15 (1993)
Nelder, J.A., Mead, R.: A simplex method for function minimization. The computer Journal 7, 308–313 (1965)

Newby, G.B.: Cognitive space and information space. Journal of the American Society for Information Science and Technology 52, 1026–1048 (2001)

Niculae, S., Leila, K., Bipin, C.D.: Using semantic templates for a natural language interface to the CINDI virtual library. Data Knowl. Eng. 55, 4–19 (2005)

Niu, L., Lu, J., Zhang, G.: Cognition-driven decision support system framework. In: The First International Conference on Risk Analysis and Crisis Response Shanghai. Atlantis Press, China (2007)

Niu, L., Lu, J., Zhang, G.: Improved business intelligence analytics on manager's experience. In: 2008 IEEE Congress on Evolutionary Computation, Hong Kong (2008)

Niu, L., Zhang, G.: A model of cognition-driven decision process for business intelligence. In: The 2008 IEE/WIC/ACM International Conference on Web Intelligence, Sydney, Australia (2008)

Noh, J.B., et al.: A case-based reasoning approach to cognitive map-driven tacit knowledge management. Expert Systems with Applications 19(4), 249–259 (2000)

O'keefe, Mceachern: Web-based customer decision support systems. ACM Transactions on Internet Technology 41, 71–78 (1998)

Ogryczak, W., Studzinski, K., Zorychta, K.: A solver for the multi-objective transshipment problem with facility location. European Journal of Operational Research 43, 53–64 (1989)

Owens, H.D., Philippakis, A.S.: Inductive consistency in knowledge-based decision support systems. Decision Support Systems 13, 167–181 (1995)

Paige, G.B., Stone, J.J., Lane, L.J., Hakonson, T.E.: Calibration and testing of simulation models for evaluation of trench cap designs. Journal of Environmental Quality 25, 136–144 (1996)

Paige, G.B., Stone, J.J., Lane, L.J., Yakowitz, D.S., Hakonson, T.E.: Evaluation of a prototype decision support system for selecting trench cap designs. Journal of Environmental Quality 25, 127–135 (1996)

Pasi, G., Yager, R.R.: Modeling the concept of majority opinion in group decision making. Information Sciences 176, 390–414 (2006)

Plous, S.: The psychology of judgment and decision making. McGraw-Hill, Humanities (1993)

Pogorelec, J.: Reverse logistics is doable, important. Frontline Solutions 1, 68–69 (1998)

Poh, K.: Knowledge-based guidance system for multi-attribute decision making. Artificial Intelligence in Engineering 12, 315–326 (1998)

Poh, K., Quaddus, M.A.: A hybrid approach to multiobjective linear optimization. Journal of Operational Research Society 41, 1037–1048 (1990)

Poh, K., Quaddus, M.A., Chin, K.L.: MOLP-PC: an interactive decision support environment for multiple objective linear optimization. In: Goh, M.C.T.L. (ed.) OR Applications in Singapore. Operational Research Society of Singapore, pp. 25–39 (1995)

Pohl, R.F.: Cognitive illusions. Psychology Press, New York (2004)

Porac, J.F., Thomas, H.: Taxonomic mental models in competitor definition. Academy of Management Review 15, 224–240 (1990)

Power, D.J.: Decision Support Systems Glossary. DSS Resources, World Wide Web (1999), `http://www.DSSResources.COM/glossary/`

Power, D.J.: Decision support systems: concepts and resources for managers. Quorum Books division Greenwood Publishing, Westport (2002)

Power, D.J., Kaparthi, S.: Building web-based decision support systems. Studies in Informatics and Control 11, 291–302 (2002)

Quaddus, M.A.: Computer aided learning of multiple criteria decision making: an integrated system and its effectiveness. Computers and Education 28, 103–111 (1997)

Quaddus, M.A., Holzman, A.G.: IMOLP: an interactive method for multiple objective linear programs. IEEE Transactions on Systems, Man, and Cybernetics SMC-16, 462–468 (1986)

Quaddus, M.A., Klass, D.: ALLOCATE: an interactive system for multi-criteria resource allocation. In: Teo, L.C.K.L., Siew, P.F., Leung, Y.H., Jennings, L.S., Rehbock, V. (eds.) Optimization Techniques and Application, Perth, Australia, pp. 471–478 (1998)

Radermacher, F.J.: Decision support systems - scope and potential. Decision Support Systems 12, 257–265 (1994)

Rangel, R.A.P., Joaquin Pérez, O., Juan Javier González, B., Gelbukh, A., Sidorov, G., Myriam, J.R.M.: A domain independent natural language interface to databases capable of processing complex queries. In: Gelbukh, A., de Albornoz, Á., Terashima-Marín, H. (eds.) MICAI 2005. LNCS (LNAI), vol. 3789, pp. 833–842. Springer, Heidelberg (2005)

Rao, H.R., Lingaraj, B.P.: Expert systems in production and operations management: classification and prospects. Interfaces 18, 80–91 (1988)

Rao, H.R., Sridhar, R., Narain, S.: An active intelligent decision support system - architecture and simulation. Decision Support Systems 12, 79–91 (1994)

Reinartz, T., Iglezakis, I., Roth-Berghofer, T.: Review and restore for case-base maintenance. Computational Intelligence 17, 214 (2001)

Reisbeck, C.K., Schank, R.C.: Inside case-based reasoning. Lawrence Erlbaum Associates, Hillsdale (1989)

Reisman, S., Johnson, T.W., Mayes, B.T.: Group decision program: a videodisc-based group decision support system. Decision Support Systems 8, 169–180 (1992)

Reiter, R.: A logic for default reasoning. Artificial Intelligence 13, 81–132 (1980)

Resnick, M.L.: Situation awareness applications to executive dashboard design. Paper presented at the Human Factors and Ergonomics Society 47th Annual Conference (2003)

Rice, D.: Reverse logistics presents opportunities. Challenges, Services News 1, 1–13 (2002)

Richardson, H.L.: Logistics in reverse. Industry Week 4/16/2001 250, 37 (2001)

Rocha, C., Schwabe, D., Aragao, M.P.: A hybrid approach for searching in the semantic web. In: Proceedings of the 13th international conference on World Wide Web, pp. 374–383 (2004)

Rogers, D.S., Lambert, D.M., Croxton, K.L., Garcia-Dastugue, S.: The returns management process. The International Journal of Logistics Management 13, 1–18 (2002)

Roth, D.: Introduction to syntactic parsing, November 18 (2004), `http://l2r.cs.uiuc.edu/~danr/Teaching/CS598-05/Lectures/Roxana.pdf` (retrieved November 20, 2007)

Rouse, W.B.: Need to know-information, knowledge, and decision making. IEEE Transactions on Systems, Man and Cybernetics, Part C: Applications and Reviews 32, 282–292 (2002)

Rouse, W.B., Morris, N.M.: On looking into the black box: Prospects and limits in the search for mental models. Georgia Institute of Technology, Center for Man-Machine Systems Research, Atlanta (1985)

Rowley, J.: The reverse supply-chain impact of current trends. Logistics and Transport Focus 2, 27–31 (2000)

Roy, R.: Problems and methods with multiple objective functions. Mathematical Programming 1, 239–266 (1971)

Ruan, D., Liu, J., Carchon, R.: Linguistic assessment approach for managing nuclear safeguards indicator information. Int. J. of Logistics Information Management 16, 401–419 (2003)

Rughooputh, H.C.S., Ah King, R.T.F.: Environmental/economic dispatch of thermal units using an elitist multi-objective evolutionary algorithm. Industrial Technology 1, 48–53 (2003)

Russo, E., Schoemaker, P.J.H.: Decision traps: The ten barriers to decision-making and how to overcome them. Simon And Schuster, New York (1990)

Saaty, T.: The analytic hierarchy process. McGraw-Hill, New York (1980)

Sage, A.P.: Decision support systems engineering. Wiley, New York (1991)

Sakawa, M.: Interactive multiobjective decision making by the sequential proxy optimization technique: SPOT. European Journal of Operational Research 9, 386–396 (1982)

Salas, E., Prince, C., Baker, D.P., Shrestha, L.: Situation awareness in team performance: implications for measurements and training. Human Factors 37, 123–136 (1995)

Salton, G., Wong, A., Yang, C.S.: A vector space model for automatic indexing. Communications of the ACM 18, 613–620 (1975)

Salukvadze, M.: On the extension of solutions in problems of optimization under vector valued criteria. Journal of Optimization Theory and Application 13, 203–217 (1974)

Sarter, N.B., Woods, D.D.: Situation awareness: a critical but ill-defined phenomenon. The International Journal of Aviation Psychology 1(1), 45–57 (1991)

Sauter, V.L.: Decision Support Systems: An Applied Managerial Approach. John Wiley, New York (1997)

Schickel-Zuber, V., Faltings, B.: OSS: a semantic similarity function based on hierarchical ontologies. Paper presented at the International Joint Conference on Artificial Intelligence, Hyderabad, India (2007)

Schmitt, N.: Naturalistic decision making in business and industrial organizations. In: Klein, G. (ed.) Naturalistic decision making. Lawrence Erlbaum Associates, Mahwah (1997)

Schubert, P., Dettling, W.: Extended web assessment method (EWAM) - evaluation of e-commerce applications from the customer's view-point. In: Proc. the 35th Annual Hawaii International Conference on System Sciences (HICSS 2002), pp. 175–184 (2002)

Schwartz, A.E.: Group decision-making. The CPA Journal 64, 60–64 (1994)

Senge, P.M.: The fifth discipline. DouleDay and Currency, New York (1990)

Shafer, G.: A Mathematical Theory of Evidence. Princeton University Press, Princeton (1976)

Shan, F., Xu, L.D.: A hybrid knowledge-based system for urban development. Expert Systems with Applications 10, 157–163 (1996)

Shim, J.P., Warkentin, M., Courtney, J.F., Power, D.J., Sharda, R., Carlsson, C.: Past, present, and future of decision support technology. Decision Support Systems 33, 111–126 (2002)

Shin, W., Ravindran, A.: Interactive multiple objective optimization: survey I - continuous case. Computers and Operations Research 18, 97–114 (1991)

Siau, K., Tan, X.: Improving the quality of conceptual modeling using cognitive mapping techniques. Data & Knowledge Engineering 55, 343–365 (2005)

SignaText, Report builder (2008), http://www.signatext.com/products.html (retrieved August 15, 2008)

Sikder, I.U., Gangopadhyay, A.: Design and implementation of a web-based collaborative spatial decision support system: organizational and managerial implications. Information Resources Management Journal 15, 33–47 (2002)

Simon, H.A.: The new science of management decision. Prentice-Hail, Englewood Cliffs (1977)

Simon, H.A.: Rational decision making in business organizations. American Economic Review 69, 493–513 (1979)

Singh, D.T.: Incorporating cognitive aids into decision support systems: the case of the strategy execution process. Decision Support Systems 24, 145–163 (1998)

Singh, S.K., Watson, H.J., Watson, R.T.: EIS support for the strategic management process. Decision Support Systems 33, 71–85 (2002)

Sol, H.G.: DSS: buzz word or OR challenge. European Journal of Operational Research 22, 1–8 (1985)

Sol, H.G.: Conflicting experiences with DSS. Decision Support Systems 3, 203–211 (1987)

Sonnenwald, D.H., Maglaughlin, K.L., Whitton, M.C.: Designing to support situation awareness across distances: an example from a scientific collaboratory. Information Processing & Management 40, 989–1011 (2004)

Spangler, W.: The role of artificial intelligence in understanding the strategic decision-making process. IEEE Transaction on Knowledge and Data Engineering 3, 145–159 (1991)

Sprague, R.H.: A framework for the development of decision support systems. Management Information System Quarterly 4, 1–26 (1980)

Sprague, R.H.: DSS in context. Decision Support Systems 3, 197–202 (1987)

Sprague, R.H., Watson, H.J.: Decision Support Systems: Putting Theory into Practice, 3rd edn. Prentice-Hall, London (1993)

Sridhar, S.: Decision support using the intranet. Decision Support Systems 23, 19–28 (1998)

Stanners, M., French, H.T.: An empirical study of the relationship between situation awareness and decision making: DSTO Systems Sciences Laboratory (2005)

Sternberg, R.J.: Cognitive psychology Belmont. Thomson/Wadsworth, CA (2006)

Steuer, R.E.: An interactive multiple objective linear programming procedure. TIMS Studies in the Management Sciences 6, 225–239 (1977)

Steuer, R.E.: Multiple Criteria Optimization: Theory, Computation, and Application. John Willy and Sons, New York (1986)

Steuer, R.E., Choo, E.U.: An interactive weighted Tchebycheff procedure for multiple objective programming. Mathematical Programming 26, 326–344 (1983)

Stewart, T.J.: A multi-criteria decision support system for R&D project selection. Journal of the Operational Research Society 42, 17–26 (1991)

Stylios, C.D., Georgopoulos, V.C., Malandraki, G.A., Chouliara, S.: Fuzzy cognitive map architectures for medical decision support systems. Applied Soft Computing 8, 1243–1251 (2008)

Sun, M., Steuer, R.E.: InterQuad: an interactive quad tree based procedure for solving the discrete alternative multiple criteria. European Journal of Operational Research 89, 462–472 (1996)

Sutcliffe, K.M., Weber, K.: The high cost of accurate knowledge. Harvard Business Review 81, 74–82 (2003)

Te'eni, D., Ginzberg, M.J.: Human-computer decision systems: the multiple roles of DSS. European Journal of Operational Research 50, 127–139 (1991)
Tecle, A., Duckstein, L.: A Procedure for Selecting MCDM Techniques for Forest Resource Management. Springer, New York (1992)
Tecle, A., Shrestha, B.P., Duckstein, L.: A multiobjective decision support system for multiresource forest management. Group Decision and Negotiation 7, 23–40 (1998)
Temperley, D., Sleator, D.: Parsing English with a link grammar. In: Third International Workshop on Parsing Technologies (1993)
Theodorson, G.A., Theodorson, A.G.: A Modern Dictionary of Sociology. Harper and Row, New York (1969)
Teodorescu, H.N., Mlynek, D., Kandel, A., Zimmermann, H.J.: Intelligent Systems and Interfaces. Springer, Boston (2000)
Travica, B., Cronin, B.: The ARgo: a strategic information system for group decision making. International Journal of Information Management 15, 223–236 (1995)
Turban, E.: Implementing decision support system: a survey. In: Proc. The 1996 IEEE International Conference on Systems, Man and Cybernetics, Beijing, China (1996)
Turban, E., Aronson, J.E.: Decision Support Systems and Intelligent Systems. Pearson Education India, London (1998)
Turban, E., Aronson, J.E., Liang, T.P.: Decision support systems and intelligent systems, 7th edn. Pearson prentice Hall, New Jersey (2005)
Uschold, M., King, M., Moralee, S., Zorgios, Y.: The enterprise ontology. The Knowledge Engineering Review 13, 31–89 (1998)
Venkatachalam, A.R., Sohl, J.E.: An intelligent model selection and forecasting system. Journal of Forecasting 18, 167–180 (1999)
Venkatesh, P., Gnanadass, R., Padhy, N.P.: Comparison and application of evolutionary programming techniques to combined economic emission dispatch with line flow constraints. IEEE Transactions on Power Systems 18, 688–697 (2003)
Vetschera, R.: MCView: an integrated graphical system to support multi-attribute decision. Decision Support Systems 11, 363–371 (1994)
Vidulich, M.A.: The role of scope as a feature of situation awareness metrics. In: International Conference on Experimental Analysis and Measurement of Situation Awareness. Embry-Riddle Aeronautical University Press, FL (1995)
Vogel, D.R., Nunamaker, J.F.: Group decision support system impact: multi-methodological exploration. Information and Management 18, 15–28 (1990)
Voogd, H.: Multicriteria Evaluation for Urban and Regional Planning, Pion, London (1983)
Voss, J.F., Post, T.A.: On the solving of ill-structured problems. In: Chi, M., Glaser, R., Farr, M. (eds.) The nature of expertise, pp. 271–295. Lawrence Erlbaum Associates, New Jersey (1988)
Walters, B.A., Jiang, J.J., Klein, G.: Strategic information and strategic decision making. In: The EIS/CEO interface in smaller manufacturing companies Information & Management, vol. 40, pp. 487–495 (2003)
Wang, H.F., Shen, S.Y.: Group decision support with MOLP applications. IEEE Transactions on Systems, Man, and Cybernatics 19, 143–153 (1989)
Wang, K.J., Chien, C.F.: Designing an internet-based group decision support system. Robotics and Computer-Integrated Manufacturing 19, 65–77 (2003)
Watson, S.R., Buede, D.M.: Decision Synthesis: The Principles and Practice of Decision Analysis. Cambridge University Press, Cambridge (1987)

Weck, M., Klocke, F., Schell, H., Rüenauver, E.: Evaluating alternative production cycles using the extended fuzzy AHP method. European Journal of Operational Research 100, 351–366 (1997)

Weistroffer, H.R.: An interactive goal-programming method for nonlinear multiple-criteria decision-making problems. Computers & Operations Research 10, 311–320 (1983)

Welter, T.R.: Tools at the top. Industry Week 237, 41–44 (1988)

Wikimedia. Regular expression (2008, July 24, 2008), `http://en.wikipedia.org/wiki/Regular_expression` (retrieved July 25, 2008)

Wong, Z., Aiken, M.: Automated facilitation of electronic meetings. Information & Management 41, 125–134 (2003)

Woods, W.A.: Progress in natural language understanding: An application to Lunar geology. In: AFIPS Conference 1973, Arlington (1973)

Yadav, S.B., Khazanchi, D.: Subjective understanding in strategic decision making: An information systems perspective. Decision Support Systems 8(1), 55–71 (1992)

Yager, R.R.: Higher structures in multi-criteria decision making. International Journal of Man-Machine Studies 36, 553–570 (1992)

Yager, R.R.: Non-numeric multi-criteria multi-person decision making. International Journal of Group Decision Making and Negotiation 2, 81–93 (1993)

Yager, R.R.: Decision making under uncertainty with ordinal information. International Journal of Uncertainty, Fuzziness and Knowledge-Based Systems 7, 483–500 (1999a)

Yager, R.R.: A game theoretic approach to decision making under uncertainty. International Journal of Intelligent Systems in Accounting, Finance and Management 8, 131–143 (1999b)

Yager, R.R.: Including decision attitude in probabilistic decision making. International Journal of Approximate Reasoning 21, 1–21 (1999c)

Yager, R.R.: On the valuation of alternatives for decision making under uncertainty. International Journal of Intelligent Systems 17, 687–707 (2002)

Yager, R.R.: A framework for multi-source data fusion. Information Sciences 163, 175–200 (2004a)

Yager, R.R.: Modeling prioritzed multi-criteria decision making. IEEE Transactions on Systems, Man and Cybernetics Part B 34, 2386–2404 (2004b)

Yager, R.R., Kikuchi, S.: On the role of anxiety in decisions under possibilistic uncertainty. IEEE Transactions on Systems, Man and Cybernetics. PartB: Cybernetics 34, 1224–1234 (2004)

Yakowitz, D.S., Hipel, K.W.: Multiple objective decision making for lokahi (balance) in environmental management. Applied Mathematics and Computation 83, 97–115 (1997)

Yakowitz, D.S., Stone, J.J., Lane, L.J., Heilman, P., Masterson, J., Abolt, J., Imam, B.: A decision support system for evaluating the effects of alternative farm management systems on water quality and economics. Water Science & Technology 28, 47–54 (1993)

Yau, C., Davis, T.: Using multi-criteria analysis for tenant selection. Decision Support Systems 12, 233–244 (1994)

Yoon, K., Hwang, C.L.: Multiple Attribute Decision Making: An Introduction. Sage Publications, London (1995)

You, Y., Pekkola, S.: Meeting others - supporting situation awareness on the WWW. Decision Support Systems 32, 71–82 (2001)

Zachary, W.: A cognitively based functional taxonomy of decision support techniques. Human-Computer Interaction 2, 25–63 (1986)

Zachary, W.: A cognitively-based functional taxonomy of decision support techniques. ACM SIGCHI Bulletin 19, 72–73 (1987)
Zadeh, L.A.: A contribution to the theory of nonlinear systems. J. Franklin Institute 255, 387–408 (1953)
Zadeh, L.A.: A theory of approximate reasoning. In: Hayes, J., Michie, D., Mikulich, L.I. (eds.) Machine Intelligence 9, pp. 149–194. Halstead Press, New York (1979b)
Zadeh, L.A.: Knowledge representation in fuzzy logic. IEEE Trans. on Knowledge and Data Engineering 1, 89–100 (1989)
Zadeh, L.A., Desoer, C.A.: Linear System Theory-The State Space Approach. McGraw-Hill Book Co., New York (1963)
Zeleny, M.: Multiple Criteria Decision Making. McGraw-Hill, New York (1982)
Zhang, W., Hill, R.: A template-based and pattern-driven approach to situation awareness and assessment in virtual humans. In: Proc. the Fourth International Conference on Autonomous Agents, Spain, pp. 116–123 (2000)
Zimmermann, H.-J.: Fuzzy sets, decision making and expert systems. Kluwer Academic Publishers, Boston (1987)
Zimmermann, H.J.: Interactive decision support for semi-structured mathematical programming models. In: Mitra, G. (ed.) Mathematical Models for Decision Support, pp. 307–319. Springer, Heidelberg (1988)
Zimmermann, H.J., Witte, E.: Empirical Research on Organisational Decision Making. North Holland, Amsterdam (1986)
Ziont, S., Wallenius, J.: An interactive programming method for solving the multiple criteria problem. Management Science 22, 652–663 (1975)

Abbreviations

ADVP	Adverb Phrase
AHP	Analytic Hierarchy Process
API	Application Programming Interface
AW	Adventure Works
AWDW	Adventure Works Data Warehouse
AWO	Adventure Works Ontology
BAM	Business Activity Monitoring
BDT	Behavioral Decision Theory
BI	Business Intelligence
CAL	Class Abstract Level
CBR	Case Based Reasoning
CCP	Context Coverage Point
CDDP	Cognition-Driven Decision Process
CDM	Classical Decision Making
CDS	Cognitive Decision Support
CIO	Chief Information Officer
CP	Context Point
CPP	Context Position Point
DBMS	Database Management System
DSS	Decision Support System
DW	Data Warehouse
EB	Experience Base
EM	Experience Management
ESS	Executive Support System
ETL	Extracting Transforming Loading
EWS	Early Warning System
GDSS	Group Decision Support System
GSS	Group Support System
GUI	Graphical User Interface
ICSP	Inverse Context Specificity Point
IDA	Intelligence Data Analysis
IDC	International Data Corporation
IDSS	Intelligent Decision Support System
IN	Information Need

IR	Information Retrieval
IS	Information System
JDM	Judgment (and) Decision Making
KN	Knowledge Need
KPIs	Key Performance Indicators
MADM	Multi Attribute Decision Making
MBMS	Model-base Management System
MCDM	Multi-Criteria Decision Making
MDR	Multidimensional Data Reporting
MDX	Multidimensional Expressions
MODSS	Multi-Objective Decision Support System
MOLAP	Multidimensional Online Analytical Processing
NDM	Naturalistic Decision Making
NLIDB	Natural Language Interface to Data Base
NLP	Natural Language Processing
NP	Noun Phrase
ODM	Organizational Decision Making
OLAP	Online Analytical Processing
OLTP	Online Transaction Processing
OM	Ontology Management
POS	Part of Speech
PP	Prepositional Phrase
Q/A	Question-Answering
QB	Query Builder
RASP	Robust Accurate Statistical Parsing
RDBMS	Relational Database Management System
RDR	Relational Data Reporting
RPD	Recognition-Primed Decision
SA	Situation Awareness
SAA	Situation Awareness Annotating
SAM	Situation Awareness Management
SAP	Situation Awareness Parsing
SD	Standard Deviation
SP	Situation Presentation
SPRINT	Strategic Plan and Resource Integration
SQL	Structured Query Language
TOPSIS	Technique for Order Preference by Similarity to Ideal Solution
UI	User Interface
VP	Verb Phrase

Index